Interactive TV Demystified

On-Line Updates

Additional updates relating to interactive TV in general, and this book in particular, can be found at the *Standard Handbook of Video and Television Engineering* web site:

www.tvhandbook.com

The tvhandbook.com web site supports the professional video community with news, updates, and product information relating to the broadcast, post production, and business/industrial applications of digital video.

Check the site regularly for news, updated chapters, and special events related to video engineering. The technologies encompassed by *Interactive TV Demystified* are changing rapidly, with new standards proposed and adopted each month. Changing market conditions and regulatory issues are adding to the rapid flow of news and information in this area. Specific services found at **www.tvhandbook.com** include:

- **Video Technology News**. News reports and technical articles on the latest developments in digital television, both in the U.S. and around the world. Check in at least once a month to see what's happening in the fast-moving area of digital television.

- **Television Handbook Resource Center**. Check for the latest information on professional and broadcast video systems. The Resource Center provides updates on implementation and standardization efforts, plus links to related web sites.

- **tvhandbook.com Update Port**. Updated material for *DTV: The Revolution in Digital Video* is posted on the site each month. Material available includes updated sections and chapters in areas of rapidly advancing technologies.

- **tvhandbook.com Book Store**. Check to find the latest books on digital video and audio technologies. Direct links to authors and publishers are provided. You can also place secure orders from our on-line bookstore.

In addition to the resources outlined above, detailed information is available on other books in the McGraw-Hill Video/Audio Series.

Interactive TV Demystified

Jerry C. Whitaker, Editor

McGraw-Hill
New York   Chicago   San Francisco
Lisbon   London   Madrid   Mexico City   Milan
New Delhi   San Juan   Seoul   Singapore
Sydney   Toronto

McGraw-Hill

A Division of The McGraw-Hill Companies

1 2 3 4 5 6 7 8 9 0 IAGM 0 9 8 7 6 5 4 3 2 1

ISBN 0-07-136325-4

The sponsoring editor for this book was Steve Chapman. The production supervisor was Pamela Pelton. The book was set in Times New Roman and Helvetica by Technical Press, Morgan Hill, CA.

Printed and bound by Quebecor/Kingsport.

McGraw-Hill books are available at special quantity discounts to use as premiums and sales promotions, or for use in corporate training programs. For more information, please write to the Director of Special Sales, McGraw-Hill, Two Penn Plaza, New York, NY 10121. Or contact your local bookstore.

This book is printed on recycled, acid-free paper containing a minimum of 50% recycled, de-inked fiber.

Dedicated to the memory of

Donald G. Fink

one of the pioneers of television engineering

Contents

Contributors

Mark Grossman, Geocast Network Systems
Scott Johnson, Teralogic
Kishore Manghnani, Teralogic
Chandy Nilakantan, SkyStream Networks
Skip Pizzi, Microsoft

Preface

The new digital television (DTV) systems now being deployed around the world hold enormous potential for new services to consumers and new revenue streams for broadcasters. In addition to greatly improved pictures and sound, the DTV systems of the Advanced Television Systems Committee (ATSC) and Digital Video Broadcasting (DVB) project have been designed to provide considerable options for systems designers in the area of data-related services.

The ATSC, for example, has developed detailed standards relating to program information and data broadcasting. Among the unique features are provisions to facilitate interactivity through the use of a dial-up (or similar) *back channel*, customization of advertisements and even programming based on location, and fully-integrated closed captioning and other text-based services. Additional efforts, now ongoing, will expand upon this foundation to fully merge the television experience with that of the Internet. This convergence of industries, long predicted, is just now beginning to take shape. The end result will be a range of features never before possible for the viewing audience.

This book is intended to address, in a broad view, the technologies involved in interactive television (ITV). The text follows a logical progression from applications to fundamental technologies to standards to implementation. Contributing authors address specific areas of these topics. The Editor gratefully acknowledges their assistance.

Every effort has been made to provide a complete and detailed explanation of the important areas of ITV, even though much of this work is ongoing. In such cases, the work completed as this book went to press is reported. As the additional details are finalized, new information will be made available on the McGraw-Hill Video/Audio Series web site, **www.tvhandbook.com**. Additional information also can be found on the author's web site, **www.technicalpress.com**.

The approach taken in *Interactive TV Demystified* is to cover the applications, basic technologies, and general implementation approaches for ITV. In a book of this size, we cannot identify all of the possible scenarios for ITV, which continues to evolve, nor is it possible to explain all of the basic technologies and implementations. Readers interested in further detail are directed to companion books in the McGraw-Hill Video/Audio Series, specifically the *Standard Handbook of Video and Television Engineering*, 3rd ed.

This publication is directed toward technical and engineering personnel involved in the design, specification, installation, and maintenance of broadcast television systems and non-broadcast professional imaging systems. The basic principles of interactive television are discussed, with emphasis on how the underlying technologies influence the ultimate applications.

The author has made every effort to cover the subject of interactive television comprehensively. Extensive references are included at the end of each chapter to direct readers to sources of additional information.

The U.S. DTV system is based—of course—on the work of the Advanced Television Systems Committee (ATSC). The ATSC has published a comprehensive set of documents that describe the DTV service. Several chapters in this book draw heavily upon the landmark work of the ATSC, and the author gratefully acknowledges this contribution to *Interactive TV Demystified*.

For those who are interested in acquiring a complete set of the ATSC documents, the applicable standards are available from the ATSC World Wide Web site (www.atsc.org). A printed version of the documents is also available for purchase from the SMPTE, White Plains, N.Y. (www.smpte.org), and the National Association of Broadcasters, Washington, D.C. (www.nab.org).

Another valuable resource is the SMPTE Television Standards on CD-ROM. This product, available for purchase from the SMPTE, contains all existing and proposed Standards, Engineering Guidelines, and Recommended Practices for television work. In this era of digital video, this product is indispensable.

It has been suggested, with some certainty, that ITV will become a key element in consumer information and entertainment. This book is intended to provide readers with a complete overview of what these exciting new technologies hold in store.

Jerry C. Whitaker
October, 2000

The Promise of Interactive Television

Jerry C. Whitaker, Editor

1.1 Introduction

The television industry is entering a new era in service to the consumer. Built around two-way interactive technologies, the, the rollout of the digital television (DTV) infrastructure opens up a new frontier in communication. Two worlds that were barely connected—television and the Internet—are now on the verge of combining into an entirely new service: namely, interactive television. Thanks to the ongoing transition of television from analog to digital, it is now possible to combine video, audio, and data within the same signal.

This combination leads to powerful new applications, with limitless possibilities of great commercial potential [1]. For example, computers can be turned into traditional TV receivers and digital set-top boxes can host applications such as interactive TV, e-commerce, and customized programming.

1.2 Interactive TV

Two terms are commonly used to describe the emerging advanced television environment: *interactive television* and *enhanced television*. For the purposes of this book, we will use the term interactive television (ITV) and define it as anything that lets a consumer engage in action with the system using a remote control or keyboard to access new and advanced services. General categories of use include the following:

- The ability to select movies for viewing at home

- E-mail and on-line chat

- News story selection and archive

- Stock market data, including personal investment portfolio performance in real-time

- Enhanced sports scores and statistics on a selective basis

- On-line real-time purchase everything from groceries to software without leaving home

There is no shortage of reasons why ITV is quickly gaining momentum— and will continue to do so as new technologies take hold. The backdrop for ITV growth comes from both the market strength of the Internet and the technology foundation that supports it. With the rapid adoption of digital video technology in the cable, satellite, and terrestrial broadcasting industries, the stage is now set for the creation of an ITV segment that meets the tests of sound economic principles and introduces to a mass consumer market a whole new range of possibilities.

For example, services will soon be available that offer interactive features for game shows, sports and other programs, interactive advertising, e-mail, and Internet access as a package deal. Rather than concentrating just on Web services, the goal is to deliver a better television experience. An important component of such services is the hard-drive-based video recorder. It is practical, with video compression, to store as much as 30 hours of television programs locally at the consumer's receiver. As the price/performance ratio of hard drives continues to move forward, even more storage will be practical. This capability opens the doors to numerous innovations and features never before possible.

Other services of the DTV era offer customized channel guides and make it easy to search for shows that subscribers want to watch, or to select future TV shows to record. Such services also make it possible for consumers to create their own customized channels filled with their favorite TV shows or custom channels that contain movies by favorite actors or directors.

Soon the entire video distribution system will handle digital media only [2]. These digital media offer content providers and consumers much higher picture quality. Unlike analog, this quality is ensured no matter how many copies are made. The digital content can be distributed in an MPEG-2 compressed format over high-bandwidth digital satellite, cable, terrestrial broadcast, Internet, or by fixed media. In the home it can be played via set-top boxes, other consumer electronic devices, or PCs, all connected by an IEEE 1394 high speed home network bus.

The potential benefits of the digital environment go beyond the picture quality. The data packets associated with digital content give the content provider access to more functions and business opportunities. For example, the video can be tagged with index pointers that allow the viewer to jump to a particular scene or to links that access supplementary information on the local disk or anywhere on the Internet. *Conditional access* methods allow the content provider to control viewing according to their business paradigms; pay-per-view can be enhanced by controlling who views and how much viewing they are entitled to, what price they pay depending on loyalty and other personal criteria, when and/or where they are, and so on.

Digital media of this type requires an investment in new infrastructures at the content provider, distributor, and consumer. This infrastructure will help content providers and distributors take advantage of the ongoing business opportunities created by the new digital media.

1.2.1 Copy Protection

With digital technology, every copy of a program is a perfect copy [2]. The digital era introduces new challenges to content producers and content providers. Their intellectual property needs to be protected so that their investments are not compromised. Also, in order for consumer electronics devices to be successful in the marketplace, they must be relatively low cost and easy to use. And yet, most consumers are not pirates and any scheme protecting the content must also take into account the perceived rights of the consumers and the ease with which they obtain, store, and continue to use the media they purchase.

1.2.2 Overview of Internet Protocols

The Internet architecture uses three core transport protocols [1]:

- *Internet Protocol* (IP)—a network protocol that defines, among other things, the structure of an IP *datagram*. A datagram has a payload and a header. The header contains a source and a destination address that has global significance. These addresses allow datagrams to be individually routed throughout the entire network on a "best-effort" basis. There is actually no guarantee at the IP layer that a datagram is correctly transmitted or that a sequence of datagrams will arrive at their destination in the same order that they were sent. Above the IP protocol, the User Datagram

Protocol and Transport Control Protocol serve to provide end-to-end functionality between a pair of host systems.

- *User Datagram Protocol* (UDP)—a service above the IP layer that provides port multiplexing and a checksum that may or may not be used. The UDP does not require the addressee to acknowledge whether or not data transmitted actually arrived properly. As such, it is not a reliable protocol, but it is still to be preferred in the context of real time streaming applications, such as video and audio, when the ability to time packet transmission against an external clock source is of much greater importance than avoiding possible data losses.

- *Transport Control Protocol* (TCP)—a service that adds more functionnalities than UDP at the price of a longer header. Along with IP, TCP guarantees a reliable, serialized, and full duplex channel. A stream of bytes generated by the sender will be passed across the Internet by TCP so that it is presented to the addressee as the same sequence of bytes and in the same order as the one generated by the sender. Unlike UDP, TCP requires a return path to send acknowledgment messages.

TCP and IP are commonly paired for networking purposes, referred to as TCP-IP.

1.2.3 Data Broadcasting

Data broadcasting (*datacasting*) offers the potential for entirely new revenue sources [3]. With modern encoders, even a high-definition (HD) broadcast requires only 16 to 18 Mbits/s of digital bandwidth, out of the 19.38 Mbits/s available in a 6 MHz broadcast band under the ATSC DTV standard. This leaves a significant amount of bandwidth that can be used for arbitrary data. Figure 1.1 illustrates one possible scenario.

There are basically two ways in which broadcasters can use this available bandwidth to generate additional revenue:

- Use the excess bandwidth to broadcast data that enhances the appeal of their TV programming and/or TV advertising in an attempt to attract more advertising dollars.

- Lease the excess bandwidth to other enterprises who want to distribute data to large numbers of users in the broadcaster's viewing area.

In practice, both of these approaches will likely be used to varying degrees by different broadcasters.

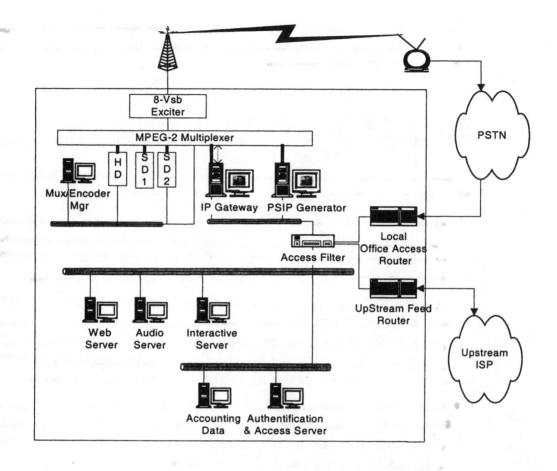

Figure 1.1 A broadcasting and datacasting terrestrial station. (*After* [1].)

For datacasting applications targeted to consumers, a key requirement for success is that large numbers of consumers have a DTV receiver that can receive and use the broadcast data. Utilizing broadcast data requires not only standards for encoding the data in the broadcast stream, but also standards for applications in the receiver to operate on the data. This could be in the form of specifications for one or more standard applications in the receiver, such as a standard HTML engine. Alternatively, it could be in the form of specifications for a standard execution environment in the receiver, so that a variety of applications could be downloaded from the broadcast stream and executed.

The ATSC technical subgroup on the DTV Application Software Environment (DASE) was working on a standard as this book went to press that incorporates both these elements.

The standardization requirement can be relaxed somewhat for applications targeted to consumers with PCs and add-in DTV cards because custom software to support such applications can be downloaded from a Web site. However, such applications are still only feasible when large numbers of consumers have DTV cards.

Datacast applications targeted to consumers can be further classified by the degree of *coupling* to the normal TV programming as follows:

- *Tightly coupled data* are intended to enhance the TV programming in real time. The viewer tunes to the TV program and receives the data enhancement along with it. In many cases the timing of the display of the broadcast data is closely synchronized with the video frames being shown.

- *Loosely coupled data* are related to the program, but are not closely synchronized with it in time. For example, an educational program might send along in the broadcast some supplementary reading materials or self-test quizzes. These might not even be viewed at the same time as the TV program. They may be saved in the DTV receiver and perused later.

- *Non-coupled data* are typically contained in separate "data-only" virtual channels. They may be data intended for real-time viewing, such as a 24-hour news headline or stock ticker service, or they may be data intended for use completely outside the DTV context.

The DASE work is discussed in more detail in Section 1.4.2.

Data Broadcast Architecture

A datacasting system must meet a number of requirements in order to properly support enterprise-to-enterprise data broadcasting [3]. One key factor in the ability of the system to meet these diverse requirements is the underlying system architecture. In particular, it should explicitly recognize and support the roles of data provider, broadcaster, and data subscriber.

The example system shown in Figure 1.2 consists of three primary components:

- *Data source server*, which allows the data provider to specify the detailed scheduling for retrieval and broadcast of individual data items. This has advantages for both the data provider and the broadcaster. The data pro-

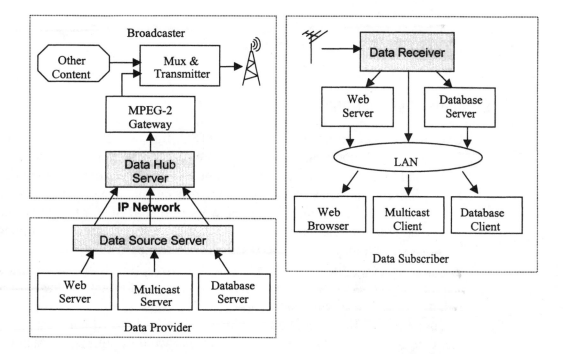

Figure 1.2 Three-component architecture for datacasting. (*After* [3].)

vider can change the scheduling at any time without involving the broadcaster in the process. The broadcaster does not have to devote any clerical resources to this task. The specifications can be transferred automatically from the data source server to the data hub server for use as needed.

- *Data hub server*, which receives data items from the data source servers and turns them over to the MPEG-2 gateway according to the schedules obtained from the data source servers. The MPEG-2 gateway simply encodes the data into MPEG-2 transport packets and sends them to a multiplexor for actual insertion into the broadcast stream.

- *Data receiver*, which extracts the data from the broadcast stream and applies decryption, decompression, and forward error recovery as needed. It provides the data subscriber with a menu of the authorized items in the broadcast stream (generated from specifications provided by the data source server and transmitted in the broadcast stream), and allows the data subscriber to select which ones to actually extract and which ones to ignore.

1.3 Computer Applications

One of the characteristics that set the ATSC digital television effort apart from the NTSC (conventional television) efforts of the past is the inclusion of a broad range of industries—not just broadcasters and receiver manufacturers, but all industries that have an interest in imaging systems. It is, of course, fair to point out that during the work of the NTSC for the black-and-white and color standards, there were no other industries involved in imaging. Be that as it may, the broad-based effort encompassed by the ATSC system has ensured that the standard will have applications in far more industries than simply broadcast television. The most visible of these allied industries is the computer business.

Computer hardware and software manufacturers lobbied hard for adjustments to the Grand Alliance DTV system that would optimize the standard for use on personal computers. Heavy-hitter vendors such as Microsoft explained that the future of television would be computers. With computers integrated into television receivers, consumers would have a host of new options and services at their fingertips, hence facilitating the public interest, convenience, and necessity.

In an industry that has seen successive waves of hype and disappointment, it is not surprising that such visions of the video future were treated skeptically, at least by broadcasters who saw these predictions by computer companies as an attempt to claim a portion of their turf.

The reality that has emerged, however, is one of cooperation among a host of industries ranging from broadcasters to cable companies to computer hardware and software manufacturers. The fact that this cooperation was forced upon the participants by market forces does not dilute the importance of the end result.

1.3.1 HDTV and Computer Graphics

Although the ATSC DTV standard is about more than just high-definition television, HDTV was—and remains—the cornerstone of the ATSC efforts (and the Grand Alliance that went before it). The core of HDTV production is the creation of high-quality images. As HDTV was emerging in the early

Table 1.1 Basic Characteristics Of IBM-Type Computer Graphics Displays (*After* [1])

Horizontal Resolution	640	800	800	1024	1280
Vertical Resolution	480	600	600	768	1024
Active Lines/Frame	480	600	600	768	1024
Total Lines/Frame	525	628	666	806	1068
Active Line Duration, μs	20.317	20.00	16.00	13.653	10.119
f_h (Hz)	37.8	37.879	48.077	56.476	76.02
f_v (MHz)	72.2	60.316	72.188	70.069	71.18
Pixel Clock, MHz	31.5	40	50	75	126.5
Video Bandwidth, MHz	15.75	20	25	37.5	63.24

1980s, a quite separate and initially unrelated explosion in electronic imaging was also under way in the form of high-resolution computer graphics. This development was propelled by broad requirements within a great variety of business and industrial applications, including:

- Computer-aided design (CAD)
- Computer-aided manufacturing (CAM)
- Printing and publishing
- Creative design (such as textiles and decorative arts)
- Scientific research
- Medical diagnosis

A natural convergence soon began between the real-time imagery of HDTV and the non-real-time high-resolution graphic systems. A wide range of high-resolution graphic display systems is commonly available today. This range addresses quite different needs for resolution within a broad spectrum of industries. Some of the more common systems are listed in Table 1.1.

The ATSC DTV system thus enjoys a good fit within an expanding hierarchy of computer graphics. This hierarchy has the range of resolutions that it does because of the varied needs of countless disparate applications. HDTV further offers an important wide-screen display organization that is eminently suited to certain critical demands. The 16:9 display, for example, can efficiently encompass two side-by-side 8 × 11-in pages, important in many print applications. The horizontal form factor also lends itself to many industrial design displays that favor a horizontally oriented rectangle, such as automobile and aircraft portrayal.

This convergence will become increasingly important in the future. The use of computer graphics within the broadcast television industry has seen enormous growth during the past decade. Apart from this trend, however, there is also the potential offered by computer graphics techniques and HDTV imagery for the creation of special effects within the motion picture production industry, already demonstrated convincingly in countless major releases.

Resolution Considerations

With few exceptions, computers were developed as stand-alone systems using proprietary display formats [4]. Until recently, computers remained isolated with little need to exchange video information with other systems. As a consequence, a variety of specific display formats were developed to meet computer industry needs that are quite different from the broadcast needs.

Among the specific computer industry needs are bright and flickerless displays of highly detailed pictures. To achieve this, computers use progressive vertical scanning with rates varying from 56 Hz to 75 Hz (or higher), referred to as *refresh rates*, and an increasingly high number of lines per picture. High vertical refresh rates and number of lines per picture result in short line durations and associated wide video bandwidths.

Because the video signals are digitally generated, certain analog resolution concepts do not directly apply to computer displays. To begin with, all *analog-to-digital* (A/D) conversion concerns related to sampling frequencies and associated anti-aliasing filters are nonexistent. Furthermore, there is no vertical resolution ambiguity, and the vertical resolution is equal to the number of active lines. The only limiting factor affecting the displayed picture resolution is the CRT dot structure and spacing, as well as the driving video amplifiers.

The computer industry uses the term *vertical resolution* when referring to the number of active lines per picture and *horizontal resolution* when referring to the number of active pixels per line. This resolution concept has no direct relationship to the television resolution concept and can be misleading (or at least confusing).

Video/Computer Optimization

In recognition of the interest on the part of the computer industry in television in general, and DTV in particular, detailed guidelines were developed by Intel and Microsoft early on in the rollout of the standard to provide for interoperability of the ATSC DTV system with personal computers of the future. Under the landmark industry guidelines known as *PC99*, design goals and interface issues for future devices and systems were addressed. To this end, the PC99 guidelines referred to existing industry standards or performance goals (benchmarks), rather than prescribing fixed hardware implementations. The video guidelines were selected for inclusion in the guide based on an evaluation of possible system and device features. Some guidelines are defined to provide clarification of available system support, or design issues specifically related to the Windows 98 and Windows NT operating system architectures (and their successor systems, Windows Millennium Edition and Windows 2000).

The requirements for digital broadcast television apply for any type of computer system that implements a digital broadcast subsystem, whether receiving satellite, cable, or terrestrial broadcasts. Such capabilities were recommended, but not required, for all system types. The capabilities were strongly recommended, however, for entertainment PC systems.

Among the specific recommendations was that systems be capable of simultaneously receiving two or more broadcast frequencies. The ability to tune to multiple frequencies results in better concurrent data and video operation. For example, with two tuners/decoders, the viewer could watch a video on one frequency and download Web pages on the other. This also enables picture-in-picture or multiple data streams on different channels or transponders. Receiver were also recommended to support conditional access mechanisms for subscription services, pay-per-view events, and other network-specific access-control mechanisms available on the broadcast services for which they were designed.

As this book went to press, the second generation of products meeting these guidelines were showing up on retail store shelves.

1.3.2 ATSC Datacasting

Although the primary focus of the ATSC system is the conveyance of entertainment programming, datacasting is a practical and viable additional feature of the standard. The concept of datacasting is not new; it has been tried with varying degrees of success for years using the NTSC system in the

U.S., and PAL and SECAM elsewhere. The tremendous data throughput capabilities of DTV, however, permit a new level of possibilities for broadcasters and cable operators.

In general, the industry has defined two major categories of datacasting [5]:

- *Enhanced television*—data content related to and synchronized with the video program content. For example, a viewer watching a home improvement program might be able to push a button on the remote control to find more information about the product being used or where to buy it.

- *Data broadcast*—data services not related to the program content. An example would be current traffic conditions, stock market activity, or even subscription services that utilize ATSC conditional access capabilities.

Effective use of datacasting could have far reaching effects on advertising and commercial broadcaster business models. A new generation of intelligent ATSC receivers with built in Internet browsers and reverse communications channels will be able to seamlessly integrate Internet services with broadcast television.

1.4   Digital Home Network

It is expected that in the near-term future, data for audio, video, telephony, printing, and control functions are all likely to be transported through the home over a digital network [6]. This network will allow the connection of devices such as computers, digital TVs, digital VCRs, digital telephones, printers, stereo systems, and remotely controlled appliances. To enable this scope of interoperability of home network devices, standards for physical layers, network information, and control protocols need to be generally agreed upon and accepted. While it would be preferable to have a single stack of technology layers, no one selection is likely to satisfy all cost, bandwidth, and mobility requirements for in-home devices.

From a broadcasting perspective, the ability to provide unrestricted entertainment services to consumer devices in the home is a key point of interest. Also, ancillary data services directed to various devices offer significant marketplace promise. Control and protocol standards to enable the delivery of selected programming from cable set-top boxes to DTV sets using IEEE 1394 have been approved by the Consumer Electronics Association (CEA) and the Society of Cable and Telecommunications Engineers (SCTE). The

CEA standard is EIA-775. The SCTE has approved two complementary standards, DVS 194 and DVS 195 (with copy protection—the *SC* system—and without copy protection).

The standards from these organizations differ in some respects, and the SCTE version has some troublesome aspects for broadcasters. For example, the SCTE standards require the cable set-top box to be the program selection device and only transfer data related to one selected program at a time to the 1394 bus. Accessing data from a broadcast stream that is unrelated to the current program selection on the cable box is also not defined in the SCTE standards.

More generally, the principal physical layer interconnections at the DTV set are expected to be RF (NTSC, VSB, QAM), baseband component (RGB, Y Pr Pb), and digital (1394). Composite video and S-video will be around a long time as well. All of these but 1394 are one-way paths.

Other appliances (or sensors) in the home may use RF on the power lines, dedicated coax, and dedicated twisted pair for control functions. From this plethora of physical layer choices, the signaling defined in the set of 1394 standards appears to be emerging as the choice for high-speed local connections and as the backbone for passing information around the home. This physical layer comes in several versions. The most mature technology is the IEEE 1394-1995 standard for communicating over 4.5 m of unshielded twisted pair (UTP). In addition, 60- and 100-m versions have been demonstrated to work over fiber. Commercial products for 1394 over plastic fiber are now available.

On top of the physical layer, network and control layers are needed. This is the part of the communications stack that was defined for the DTV interface (i.e., EIA-775), and was being defined at this writing for long distance 1394 by the Video Electronics Standards Association (VESA) Home Network group (VHN). Their objective was to define a network layer approach to allow seamless operation across different physical layers using Internet protocol (IP). HAVI (Home Audio/Video Interoperability) is a different network layer solution for home networking that has the same unifying objective as VHN. Optimized for the 1394 physical layer, HAVI is focused on audio/video applications and requirements.

On top of the network layer, widely differing approaches exist to the application interfaces, operating system, rendering engines, browsers, and the degree of linkage with the Internet.

1.4.1 Advanced Television Enhancement Forum

The Advanced Television Enhancement Forum (ATVEF) is a cross-industry group formed to specify a single public standard for delivering interactive television experiences that can be authored once—using a variety of tools—and deployed to a variety of television, set-top, and PC-based receivers [7]. Version 1.1 is a foundation specification, defining the fundamentals necessary to enable creation of HTML-enhanced television content so that it can be reliably broadcast across any network to any compliant receiver.

The ATVEF specification for enhanced television programming uses existing Internet technologies. It delivers enhanced TV programming over both analog and digital video systems using terrestrial, cable, satellite, and Internet networks. The specification can be used with both one-way broadcast and two-way video systems, and is designed to be compatible with all international standards for both analog and digital video systems.

The ATVEF specification consists of three principal parts:

- Content specifications to establish minimum requirements for receivers

- Delivery specifications for transport of enhanced TV content

- A set of specific bindings

A central design point of the ATVEF document was to use existing standards wherever possible and to minimize the creation of new specifications. The content creators in the group determined that existing Web standards, with only minimal extensions for television integration, provide a rich set of capabilities for building enhanced TV content in today's marketplace. The ATVEF specification references full existing specifications for HTML, ECMAScript, DOM, CSS, and media types as the basis of the content specification. The guidelines are not a limit on what content can be sent, but rather are intended to provide a common set of capabilities so that content developers can author content once and reproduce it on a wide variety of players.

Another key design goal was to provide a single solution that would work on a wide variety of networks. ATVEF is capable of running on both analog and digital video systems as well as networks with no video capabilities at all. The specification further supports transmission across terrestrial broadcast, cable, and satellite systems, and the Internet. In addition, it will bridge between networks; for example, in a compliant system, data on an analog terrestrial broadcast will bridge to a digital cable system. This design goal was achieved through the definition of a transport-independent content format and the use of IP as the reference binding. Because IP bindings already exist for each of these video systems, ATVEF can take advantage of a wealth of previous work.

The specification defines two transports—one for broadcast data and one for data pulled through a return path. While the ATVEF specification has the capability to run on any video network, a complete specification requires a specific binding to each video network standard in order to ensure true interoperability.

Reference and example bindings also are specified in the document, although it is assumed that appropriate standards bodies will define the bindings for each video standard—PAL, SECAM, DVB, ATSC, and others.

There are many roles in the production and delivery of television enhancements. The ATFEF document identifies three key roles:

- **Content creator.** The content creator originates the content components of the enhancement including graphics, layout, interaction, and triggers.

- **Transport operator.** The transport operator runs a video delivery infrastructure (terrestrial, cable, or satellite) that includes a transport mechanism for ATVEF data.

- **Receiver.** The receiver is a hardware and software implementation (television, set-top box, or personal computer) that decodes and plays ATVEF content.

A particular group or company may participate as one, two or all three of these roles. (The ATVEF effort is discussed in more detail in Chapter 7.)

1.4.2 Digital Application Software Environment

The ATSC T3/S17 specialist group was charged with defining a *digital television application software environment* (DASE) for broadcast interactive applications. This environment contains two principal components: a *presentation engine* for declarative applications, and a Java-based set of interfaces for procedural applications [8].

The purpose of the presentation engine is to integrate so-called *declarative content* with streaming audio and video, and to deliver the resulting content to the television display. In partnership with the presentation engine is a set of Java interfaces that provide a means for content authors to develop procedural applications for drawing to the screen.

In defining the presentation engine, there are two distinct interfaces: the *content authoring* specification and the receiver specification. This split approach provides two benefits. First, it allows for content authoring tools to mature at a different rate than the installed base of client receivers. Second, it provides a means for authors to develop content that is delivered to different platforms with different capabilities (such as a television set-top box and a mobile phone).

The requirement for supporting different receiver profiles is based on the premise that receivers may be classified based on features that are discernible to the customer, support multiple price-point strategies, and are simple to implement and simple to understand. The presentation engine supports these requirements through modularization of the features, a layering scheme for delivering these features, and support for backward compatibility of content.

Predictive Rendering

Traditionally, television-based content producers have a great deal of control over how their product appears to the customer [9]. The conventional television paradigm would be problematic if there were no assurances that a content producer could predict how their content would be rendered on every receiver-display combination in use by consumers. The requirement for *predictive rendering*, then, is essentially a contract between the content developer and the receiver manufacturer that guarantees the following parameters:

- Content will be displayed at a specific time

- Content will be displayed in a specific sequence

- Content will look a certain way

The presentation engine supports these requirements through a well-defined model of operation, media synchronization, pixel-level positioning, and the fact that it is a *conformance specification*. The model of operation formally defines the relationship between broadcast applications, native applications, television programs, and on-screen display resources.

Pixel level positioning allows a content author to specify where elements are rendered on a display. It also allows content authors to specify elements in relation to each other or relative to the dimensions of the screen.

The presentation engine architecture consists of five principal components:

- *Markup language,* which specifies the content of the document

- *Style language,* which specifies how the content is presented to the user

- *Event model,* which specifies the relationship of events with elements in the document

- *Application programming interfaces,* which provide a means for external programs to manipulate the document

- *Media types,* which are simply those media formats that require support in a compliant receiver

1.5 DTV Product Classification

In the analog days, the interconnection of a television set with a peripheral device was a relatively minor task. In the digital era of today, however, understanding what devices will work together is not a minor consideration. For this reason, the Consumer Electronics Association announced new definitions and labels for DTV products. The definitions were expected to be incorporated into manufacturers' television marketing materials as DTV receivers continued into the retail channels.

The CEA Video Division Board resolved that analog-only televisions (televisions/monitors with a scanning frequency of 15.75 kHz) should not be marketed or designated to consumers as having any particular DTV capabilities or attributes. In a second related resolution, the Board agreed that the new definitions for monitors and tuners should be used by all manufacturers and retailers to replace general, non-industry terminology like "DTV-ready" or "HDTV-ready." They also defined minimums for HDTV displays as those with active top-to-bottom scan lines of 720 progressive or 1080 interlaced, or higher. Manufacturers were also to disclose the number of active scan lines for a high-definition image within a 16:9 aspect ratio "letter boxed" image area if the unit has a 4:3 HDTV display.

The CEA digital television definitions are as follows [10]:

- **High-definition television** (HDTV): a complete product/system with the following minimum performance attributes: 1) receives ATSC terrestrial digital transmissions and decodes all basic ATSC video formats (commonly referred to as Table 3 formats); 2) has active vertical scanning lines of 720P, 1080I, or higher; 3) capable of displaying a 16:9 image; and 4) receives and reproduces, and/or outputs Dolby Digital audio.

- **High-definition television monitor**: a monitor or display with the following minimum performance attributes: 1) has active vertical scanning lines of 720P, 1080I, or higher; and 2) capable of displaying a 16:9 image. In specifications found on product literature and in owner's manuals, manufacturers were required to disclose the number of vertical scanning lines in the 16:9 viewable area, which must be 540P, 1080I or higher to meet the definition of HDTV.

- **High-definition television tuner:** an RF receiver with the following minimum performance attributes: 1) receives ATSC terrestrial digital transmissions and decodes all ATSC Table 3 video formats; 2) outputs the ATSC Table 3 720P and 1080P/I formats in the form of HD with minimum active vertical scanning lines of 720P, 1080I, or higher, and 3): receives and reproduces, and/or outputs Dolby Digital audio. Additionally, this tuner may output HD formats converted to other formats. The lower resolution ATSC Table 3 formats can be output at lower resolution levels. Alternatively, the output can be a digital bitstream with the full resolution of the broadcast signal.

- **Enhanced definition television** (EDTV): a complete product/system with the following minimum performance attributes: 1) receives ATSC terrestrial digital transmissions and decodes all ATSC Table 3 video formats; 2) has active vertical scanning lines of 480P or higher; and 3) receives and reproduces, and/or outputs Dolby Digital audio. The aspect ratio is not specified.

- **Enhanced definition television monitor**: a monitor or display with the following minimum performance attributes: 1) has active vertical scanning lines of 480P or higher. No aspect ratio is specified.

- **Enhanced definition television tuner**: an RF receiver with the following minimum performance attributes: 1) receives ATSC terrestrial digital transmissions and decodes all ATSC Table 3 video formats; 2) outputs the ATSC Table 3 720P and 1080I/P and 480P formats with minimum active vertical scanning lines of 480P; and 3) receives and reproduces, and/or outputs Dolby Digital audio. Alternatively, the output can be a digital bitstream output capable of transporting 480P, except the ATSC Table 3 480I format, which can be output at 480I.

- **Standard definition television** (SDTV): a complete product/system with the following performance attributes: 1) receives ATSC terrestrial digital transmissions and decodes all ATSC Table 3 video formats, and produces a useable picture; 2) has active vertical scanning lines less than that of EDTV; and 3) receives and reproduces usable audio. No aspect ratio is specified.

- **Standard definition television tuner**: an RF receiver with the following minimum performance attributes: 1) receives ATSC terrestrial digital transmissions and decodes all ATSC Table 3 video formats; 2) outputs all ATSC table 3 formats in the form of NTSC output; and 3) receives and reproduces, and/or outputs Dolby Digital audio.

These industry standard definitions were intended to eliminate the confusion over product features and capabilities of television sets and monitors intended for DTV applications. The agreement promised to spur the sale of DTV-compliant sets by injecting a certain amount of logic into the marketing efforts of TV set manufacturers. The consumer electronics industry had come under fire early in their DTV product rollouts because of the use of confusing and—in many cases meaningless—marketing terms. For example, terms such as "DTV-ready" means different things to different people.

1.5.1 Cable/DTV Receiver Labeling

On the heels of the CEA DTV product classification agreement, the FCC adopted a Report and Order requiring standardized labeling of DTV receivers that are marketed for connection to cable television systems. The order specified that such receivers offered for sale after July 1, 2001, must be permanently marked with a label on the outside of the product that reads: "Meets FCC Labeling Standard Digital Cable Ready (DCR) x," where $x = 1$, 2, or 3 [11].

The Commission prohibited marketing of receiving devices claimed to be fully compatible with digital cable services unless they have the functionality of one of the three categories defined. The new rules cover any consumer television receiving device with digital signal processing capability that is intended to be used with cable systems. The rules permit marketing devices with less capability, provided full compatibility is not claimed.

DCR 1 refers to a consumer electronics television receiving device capable of receiving analog basic, digital basic, and digital premium cable television programming by direct connection to a cable system providing digital programming. This device does not have an IEEE 1394 connector or other digital interlace. A security card or point of deployment module provided by the cable operator is required to view encrypted programming.

DCR 2, a superset of DCR 1, adds a 1394 digital interface connector. It is clear from the FCC order that other digital interfaces also could be present. The FCC noted that connection of a DCR 2 receiver to a digital set-top box (also presumably a level 2 device) may support advanced and interactive digital services and programming delivered by the cable system to the set-top box.

The distinction asserted for DCR 3 is the addition of the capability to receive advanced and interactive digital services. A device with this label is not required to have a 1394 connector. As such services are not defined, the meaning of this distinction is unclear. The FCC did state that additional industry work was required to design specifications for the DCR 3 category of receivers, and that it would keep the record open in this proceeding, allowing the option to incorporate these specifications into the rules at a later date.

The FCC required the consumer electronics and cable industries to report their progress on developing technical standards in two other areas: direct connection of digital TV receivers to digital cable television systems and the provision of tuning and program schedule information to support on-screen program guides for consumers. The FCC said these two issues had been substantially, but not completely, resolved in an earlier agreement. Reporting requirements were consolidated into a single reporting timetable that began on October 31, 2000, and every six months thereafter until October 2002.

When these rules were announced by the Commission, it was unclear how much the Report and Order would aid in the delivery of DTV over a cable system to a DTV receiver, other than preventing some grossly misleading marketing practices. The actual connection of the cable systems coaxial cable to the DTV set is not covered. Presumably the products will use the relevant CEA or Society of Cable Television Engineers standards for the coax interface, which were being harmonized as this book went to press. The FCC also stopped short of selecting the standard for transport of signals over the IEEE-1394 physical interface (technically only requiring a connector, not any signaling through this connector). It is unlikely that a manufacturer will put a 1394 connector on a product without some implementation of a protocol using one of the 1394-based protocol standards. Unfortunately, TV receivers with the DCR 2 label are not assured to work with set-top boxes with the same label because of multiple 1394-based protocols.

1.6   Characteristics of the Video Signal

High-definition television has improved on earlier techniques primarily by calling more fully upon the resources of human vision [12]. The primary objective of HDTV has been to enlarge the visual field occupied by the video image, an attribute that can be used to benefit interactive TV applications. This attribute has called for larger, wider pictures that are intended to be viewed more closely than conventional video. To satisfy the viewer upon this closer inspection, the HDTV image must possess proportionately finer detail and sharper outlines.

1.6.1   Critical Implications for the Viewer and Program Producer

In its search for a "new viewing experience," early experiments conducted an extensive psychophysical research program in which a large number of attributes were studied. Viewers with nontechnical backgrounds were exposed to a variety of electronic images, whose many parameters were then varied over a wide range. A definition of those imaging parameters was being sought, the aggregate of which would satisfy the average viewer that the TV image portrayal produced an emotional stimulation similar to that of the large-screen film cinema experience.

Central to this effort was the pivotal fact that the image portrayed would be large—considerably larger than current NTSC television receivers. Some of the key definitions being sought by researchers were precisely how large, how wide, how much resolution, and the optimum viewing distance of this new video image.

A substantial body of research gathered over the years has established that the average U.S. consumer views the TV receiver from a distance of approximately seven picture heights. This translates to perhaps a 27-in NTSC screen viewed from a distance of about 10 ft. At this viewing distance, most of the NTSC artifacts are essentially invisible, with perhaps the exception of cross color. Certainly the scanning lines are invisible. The luminance resolution is satisfactory on camera close-ups. A facial close-up on a modern high-performance 525-line NTSC receiver, viewed from a distance of 10 ft, is quite a realistic and pleasing portrayal. But the system quickly fails on many counts when dealing with more complex scene content.

Wide-angle shots (such as jersey numbers on football players) are one simple and familiar example. TV camera operators, however, have adapted to this inherent restriction of 525-line NTSC, as witnessed by the continual

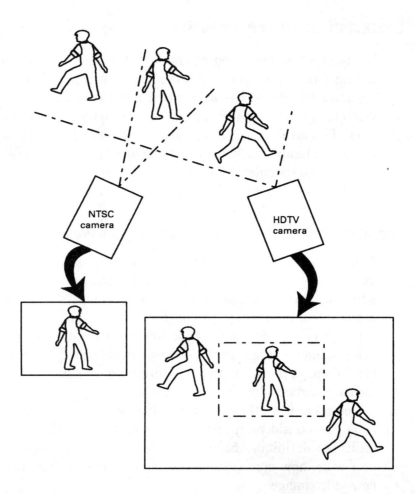

Figure 1.3 An illustration of the differences in the scene capture capabilities of conventional video and HDTV. (*After* [9].)

zooming in for close-ups during most sporting events. The camera operator accommodates for the technical shortcomings of the conventional television system and delivers an image that meets the capabilities of NTSC, PAL, and SECAM quite reasonably. There is a penalty, however, as illustrated in Figure 1.3. The average home viewer is presented with a very narrow angle of view—on the order of 10°. The video image has been rendered "clean" of many inherent disturbances by the 10-ft viewing distance and made adequate in resolution by the action of the camera operator; but, in the process, the scene has become a small "window". The now "acceptable" television image pales in comparison with the sometimes awesome visual stimulation of the cinema. The primary limitation of conventional TV systems is, therefore, image size. A direct consequence is further limitation of image content; the angle of view constantly is constricted by the need to provide adequate resolution. There is significant, necessary, and unseen intervention by the TV program director in the establishment of image content that can be passed on to the home viewer with acceptable resolution.

Compared with the 525-line NTSC signal (or the marginally better PAL and SECAM systems), the ATSC DTV system and the North American HDTV studio standard (SMPTE[1] 240M) and its digital representation (SMPTE 274M) offer a vast increase in total information contained within the visual image. If all this information is portrayed on an appropriate HDTV studio monitor (commonly available in 19-, 28-, and 38-in diagonal sizes), the dramatic technical superiority of HDTV over conventional technology easily can be seen. The additional visual information, coupled with the elimination of composite video artifacts, portrays an image almost totally free (subjectively) of visible distortions, even when viewed at a close distance.

On a direct-view CRT monitor, HDTV displays a technically superb picture. The *information density* is high; the picture has a startling clarity. However, when viewed from a distance of approximately seven picture heights, it is virtually indistinguishable from a good NTSC portrayal. The wider aspect ratio is the most dramatic change in the viewing experience at normal viewing distances.

1.6.2 Image Size

If HDTV is to find a home with the consumer, it will find it in the living room. If consumers are to retain the average viewing distance of 10 ft, then the minimum image size required for an HDTV screen for the average definition of a totally new viewing experience is about a 75-in diagonal. This represents an image area considerably in excess of present "large" 27-in NTSC (and PAL/SECAM) TV receivers. In fact, as indicated in Figure 1.4, the viewing geometry translates into a viewing angle close to 30° and a distance of only three picture heights between the viewer and the HDTV screen.

---

1. Society of Motion Picture and Television Engineers, the leading video production standards organization.

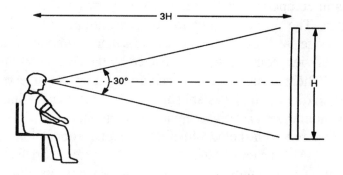

Figure 1.4 Viewing angle as a function of screen distance for HDTV. (*After* [8].)

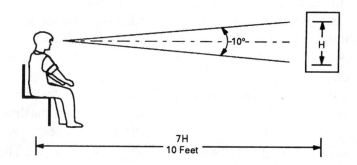

Figure 1.5 Viewing angle as a function of screen distance for conventional video systems. (*After* [12].)

Compare this with the viewing angle for conventional systems at 10°, as shown in Figure 1.5.

HDTV Image Content

There is more to the enhanced viewing experience than merely increasing picture size [12]. Unfortunately, this fundamental premise has been ignored in some audience surveys. The larger artifact-free imaging capability of HDTV allows a new image portrayal that capitalizes on the attributes of the larger screen. As mentioned previously, as long as the camera operator appropriately fills the 525 (or 625) scanning system, the resulting image (from a resolution viewpoint) is actually quite satisfactory on conventional systems. If, however, the same football game is shot with an HDTV camera the angle of view of the lens is adjusted to portray the same resolution (in the picture center) as the 525 camera when capturing a close-up of a player on its 525 screen, a vital difference between the two pictures emerges: the larger HDTV image contains considerably more information, as illustrated in Figure 1.3.

The HDTV picture shows more of the football field—more players, more of the total action. Thus, the HDTV image is radically different from the NTSC portrayal. The individual players are portrayed with the same resolution on the retina—at the same viewing distance—but a totally different viewing experience is provided for the consumer. The essence of HDTV imaging is this greater sensation of reality.

The real, dramatic impact of HDTV on the consumer will be realized only when two key ingredients are included:

- Presentation of an image size of approximately 75 in diagonal (minimum).

- Presentation of image content that capitalizes on new camera freedom in formatting larger, wider, and more true-to-life angles of view.

1.6.3 Format Development

Established procedures in the program production community provide for the 4:3 aspect ratio of video productions and motion picture films shot specifically for video distribution. This format convention has, by and large, been adopted by the computer industry for desktop computer systems.

In the staging of motion picture films intended for theatrical distribution, no provision generally is made for the limitations of conventional video displays. Instead, the full screen, in wide aspect ratios—such as CinemaScope—is used by directors for maximum dramatic and sensory impact. Consequently, cropping of essential information may be encountered more often than not on the video screen This problem is particularly acute in wide-screen features where cropping of the sides of the film frame is necessary to produce a print for video transmission. This objective is met in one of the following ways:

- *Letter-box* transmission with blank areas above and below the wide-screen frame. Audiences in North America and Japan have not generally accepted this presentation format, primarily because of the reduced size of the picture images and the aesthetic distraction of the blank screen areas.

- Printing the full frame height and cropping equal portions of the left and right sides to provide a 4:3 aspect ratio. This process frequently is less than ideal because, depending upon the scene, important visual elements may be eliminated.

- Programming the horizontal placement of a 4:3 aperture to follow the essential picture information. Called *pan and scan*, this process is used in producing a print or in making a film-to-tape transfer for video viewing. Editorial judgment is required for determining the scanning cues for horizontal positioning and, if panning is used, the rate of horizontal movement. This is an expensive and laborious procedure and, at best, it compromises the artistic judgments made by the director and the cinematographer in staging and shooting, and by the film editor in postproduction.

These considerations are also of importance to the computer industry, which is keenly interested in multimedia technology.

One of the reasons for moving to a 16:9 format is to take advantage of consumer acceptance of the 16:9 aspect ratio commonly found in motion picture films. Actually, however, motion pictures are produced in several formats, including:

- 4:3 (1.33)

- 2.35, used for 35 mm anamorphic CinemaScope film

- 2.2 in a 70 mm format

Still, the 16:9 aspect ratio generally is supported by the motion picture industry. Figure 1.6 illustrates some of the more common aspect ratios.

The SMPTE has addressed the mapping of pictures in various aspect ratios to 16:9 in Recommended Practice 199-1999. The Practice describes a method of mapping images originating in aspect ratios different from 16:9 into a 16:9 scanning structure in a manner that retains the original aspect ratio of the work. Ratios of 1.33 to 2.39 are described in RP199-1999 [14].

1.6.4 Case Histories

It has long been stated that the great benefits of both the higher resolution offered by HDTV and the wider aspect ratio will remake the way sporting events are covered. The CBS Television Network tested that theory in a series of football broadcast, beginning with the Buffalo versus New York

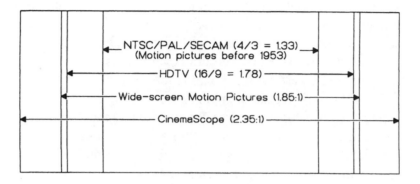

Figure 1.6 Comparison of the aspect ratios of television and motion pictures. (*After* [13].)

game on November 8, 1998. That event marked the first professional football contest to be broadcast live in the digital 1920 × 1080 HDTV format.

The historic telecast was the first of four professional football games presented in HD by CBS Sports during the season. The live HDTV telecasts were produced and transmitted independent of the regularly scheduled CBS Sports football coverage being broadcast on the conventional analog network.

Other pioneering efforts included Super Bowl XXXIV and the 1999/2000 season of "Monday Night Football" (MNF), which were broadcast live in high-definition television at 720 progressive (720P), ABC's selected HDTV format. The 1999/2000 season of MNF was the first live, regularly scheduled HDTV sporting event in primetime. The HDTV telecasts were produced and transmitted independent of ABC's "Monday Night Football" on the traditional analog network.

1.7 Aural Component of Visual Realism

The realism of a video presentation depends to a great degree on the realism of the accompanying sounds. This important point should not be lost on the designers of interactive TV systems. Particularly in the close viewing of HDTV images, if the audio system is monophonic, the sounds seem to be confined to the center of the screen. The result is that the visual and aural senses convey conflicting information. From the beginning of HDTV system design, it has been clear that stereophonic sound must be used. The generally accepted quality standard for high-fidelity audio has been set by the digital compact disc (CD). This medium covers audio frequencies from below 30 Hz to above 20 kHz, with a dynamic range of 90 dB or greater.

Sound is an important element in the viewing environment. To provide the greatest realism for the viewer, the picture and the sound should be complementary, both technically and editorially. The sound system should match the picture in terms of positional information and offer the producer the opportunity to use the spatial field creatively. The sound field can be used effectively to enlarge the picture. A *surround sound* system can further enhance the viewing experience.

1.7.1 Hearing Perception[1]

Human hearing is quite sensitive in some respects and less so in others. In the past few decades, psychoacoustic research has discovered and explored several areas in which human hearing exhibits such reduced sensitivity. This knowledge has been applied to digital encoding systems to achieve greater coding efficiencies for audio signals. Unlike standard file-based data compression systems that analyze a bit stream for statistical redundancies and remove them in a fully retroactive way (so-called *lossless* coding, such as that used in various "zip" utilities), so-called *perceptual coding* systems analyze digital audio signals and re-encode them at lower bit rates with generally inaudible results, due to their exploitation of the areas of human hearing that are relatively non-discriminating. This encoding is *lossy* in that once it is performed the audio content cannot be fully reconstructed to its original form, but ideally, the listener cannot tell the difference.

The use of perceptual coding has become commonplace in today's digital audio systems. An example is the popular MP3 format, named after a perceptual coding format for digital audio signals (MPEG-1 Audio Layer 3). To understand the value of these systems, consider the fundamental uncompressed format for digital audio coding, linear *pulse code modulation* (PCM). This format necessarily observes the Nyquist value for sampling of the digital signal (i.e., sampling at twice the frequency of the highest audio frequency intended to be reproduced on the system, or higher). For the full-range audio used in broadcast television, this implies an audio bandwidth of

---

1. This section contributed by Skip Pizzi, Microsoft

approximately 20 kHz, requiring a sampling rate of at least 40 kHz. Typical sampling frequencies used are 44.1 kHz (the audio CD sampling rate) or 48 kHz (common in digital television systems). For high-fidelity dynamic range, a resolution of 16 to 24 bits must be applied to each sample. This results in a digital audio coding that produces approximately 1 Mbits/s of data per audio channel. For the multichannel audio systems used in DTV or cinematic productions, this can amount to over 5 Mb/s or nearly 40 MB per hour of audio.

Perceptual coding systems can provide substantial reductions in this data rate—on the order of 10—without significant audio quality loss. This requires substantially less bandwidth for transmission, and correspondingly less storage space, hence the popularity of such data compression systems. They perform this reduction in data rate by adaptively reducing the resolution of groups of audio samples, while leaving the sampling rate unmodified. In this way, the audio bandwidth remains unchanged, but the dynamic range is reduced (i.e., the signal becomes temporarily noisier).

The perceptual coding systems can get away with this added noise because their operation exploits the *masking* properties of human hearing. One form of this phenomenon, *spectral masking*, dictates that in the presence of a prominent tone, other slightly lower level tones at nearby frequencies are rendered inaudible. This means that the loudest tones in a given moment of sound produce a selective desensitizing of the hearing sense to other nearby frequencies. Another form of the process, called *temporal masking*, acknowledges that a loud aural event desensitizes the hearing sense to other, quieter sounds immediately after (and even slightly before) the loud sound.

Perceptual coding systems use these *selective desensitivities* to their advantage by placing the noise products that result from their resolution reductions into the desensitized zones, thereby making the added noise inaudible. Through the use of these techniques, relatively high-fidelity audio is possible at data rates of less than 50 kbits/s per channel. This data rate would produce a dynamic range of about 10 dB (an unlistenably noisy signal) using linear PCM coding, while with perceptual coding this same number of bits can offer the equivalent of a 90+ dB dynamic range (very high fidelity).

In multichannel systems, further reductions in data rate can be taken because similar audio signals often occupy more than one of the audio channels at any given moment. In other words, it is extremely unlikely that all six channels of a surround sound signal will contain completely discrete signals having no common information. When such common information does occur, it need not be uniquely coded for each channel. This technique is often referred to as *joint coding*, and its efficiency is utilized by the systems employed in today's digital multichannel sound-for-picture formats, such as Dolby AC-3 and DTS.

1.7.2 The Aural Image

There is a large body of scientific knowledge on how humans localize sound. Most of the research has been conducted with subjects using earphones to listen to monophonic signals to study *lateralization*. *Localization* in stereophonic listening with loudspeakers is not as well understood, but the research shows the dominant influence of two factors: *interaural amplitude* differences and *interaural time delay*. Of these two properties, time delay is the more influential factor. Over intervals related to the time it takes for a sound wave to travel around the head from one ear to the other, interaural time clues determine where a listener will perceive the location of sounds. Interaural amplitude differences have a lesser influence. An amplitude effect is simulated in stereo music systems by the action of the stereo balance control, which adjusts the relative gain of the left and right channels. It is also possible to implement stereo balance controls based on time delays, but the required circuitry is more complex.

A listener positioned along the line of symmetry between two loudspeakers will hear the center audio as a phantom or *virtual image* at the center of the stereo stage. Under such conditions, sounds—dialogue, for example— will be spatially coincident with the on-screen image. Unfortunately, this coincidence is lost if the listener is not positioned properly with respect to the loudspeakers. Figure 1.7 illustrates the sensitivity of listener positioning to aural image shift. As illustrated, if the loudspeakers are placed 6 ft apart with the listener positioned 10 ft back from the speakers, an image shift will occur if the listener changes position (relative to the centerline of the speakers) by just 16 in. The data shown in the figure is approximate and will yield different results for different types and sizes of speakers. Also, the effects of room reverberation are not factored into the data. Still, the sensitivity of listener positioning can be seen clearly. Listener positioning is most critical when the loudspeakers are spaced widely, and less critical when they are spaced closely. To limit loudspeaker spacing, however, runs counter to the purpose of wide-screen displays. The best solution is to add a third audio channel dedicated exclusively to the transmission of center-channel signals for reproduction by a center loudspeaker positioned at the video display, and

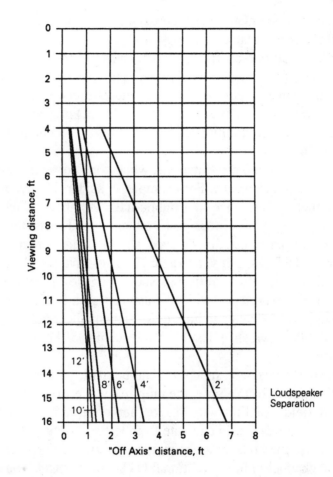

Figure 1.7 The effects of listener positioning on center image shift. (*After* [15].)

to place left and right speakers apart from the display to emphasize the wide-screen effect. The addition of *surround sound* speakers further improves the realism of the aural component of the production.

1.7.3 Matching Audio to Video

It has been demonstrated that even with no picture to provide visual cues, the ear/brain combination is sensitive to the direction of sound, particularly in an arc in front of and immediately in back of the listener. Even at the sides, listeners are able to locate direction cues with reasonable accuracy. With a large-screen display, visual cues make the accuracy of sound positioning even more important.

If the number of frontal loudspeakers and the associated channels is increased, the acceptable viewing/listening area can be enlarged. Three-channel frontal sound using three loudspeakers provides good stereo listening for three or four viewers, and a 4-channel presentation increases the area even more. The addition of one or more rear channels permits surround sound effects.

Surround sound presentations, when done correctly, significantly improve the viewing experience. For example, consider the presentation of a concert or similar performance in a public hall. Members of the audience, in addition to hearing the direct performance sound from the stage, also receive reflected sound, usually delayed slightly and perhaps diffused, from the building surfaces. These acoustic elements give a hall its tonal quality. If the spatial quality of the reflected sound can be made available to the home viewer, the experience will be enhanced greatly. The home viewer will see the stage performance in high definition and hear both the direct and indirect sound, all of which will add to the feeling of being present at the performance.

In sports coverage, much use can be made of positional information. In a tennis match, for example, the umpire's voice would be located in the center sound field—in line with his or her observed position—and crowd and ambient sounds would emanate from left and right.

Several methods have been used to successfully convey the surround sound channel(s) in conventional NTSC broadcasts. The Dolby AC-3 sound system is used in the ATSC DTV system, offering 5.1 channels of audio information to accompany the HDTV image.

1.7.4 Making the Most of Audio

In any video production, there is a great deal of sensitivity to the power of the visual image portrayed through elements such as special effects, acting, and directing that build the scene. All too often, however, audio tends to becomes separated from the visual element. Achieving a good audio product is difficult because of its subjective content. There are subtleties in the visual area, understood and manipulated by video specialists, that an audio specialist might not be aware of. By the same token, there are psychoacoustic subtleties relating to how humans hear and experience the world around them that audio specialists can manipulate to their advantage.

Reverb, for example, is poorly understood; it is more than just echo. This tool can be used creatively to trigger certain psychoacoustic responses in an audience. The brain will perceive a voice containing some reverb to be louder. Echo has been used for years to effectively change positions and dimensions in audio mixes.

To use such psychoacoustic tools is to work in a delicate and specialized area, and audio is a subjective discipline that is short on absolute answers. One of the reasons it is difficult to achieve good quality sound is because it is hard to define what that is. It is usually easier to quantify video than audio. Most people, given the same video image, come away with the same perception of it. With audio, however, accord is not so easy to come by. Musical instruments, for example, are harmonically rich and distinctive devices. A violin is not a pure tone; it is a complex balance of textures and harmonics. Audio offers an incredible palette, and it is acceptable to be different. Most video images have any number of absolute references by which images can be judged. These references, by and large, do not exist in audio.

When an audience is experiencing a program—be it a television show or an aircraft simulator computer game—there is a balance of aural and visual cues. If the production is done right, the audience will be drawn into the program, putting themselves into the events occurring on the screen. This *suspension of disbelief* is the key to effectively reaching the audience.

1.7.5 Ideal Sound System

Based on the experience of the film industry, it is clear that for the greatest impact, HDTV sound should incorporate, at minimum, a 4-channel system with a center channel and surround sound. Figure 1.8 illustrates the optimum speaker placement for enhancement of the viewing experience. This viewpoint was taken into consideration by the ATSC in its study of the Grand Alliance system.

1.7.6 Dolby AC-3

Under the ATSC DTV sound system, complete audio programs are assembled at the user's receiver from various services sent by the broadcaster. The concept of assembling services at the user's end was intended to provide for greater flexibility, including various-language multichannel principal programs supplemented with optional services for the hearing impaired and visually impaired.

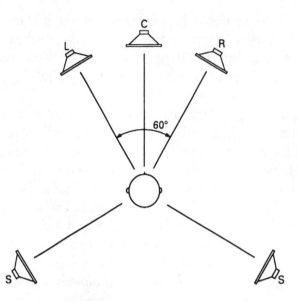

Figure 1.8 Optimum system speaker placement for HDTV viewing. (*After* [16].)

A variety of multichannel formats for the main audio services also is provided, adapting program by program to the best stereo presentation for a particular offering. An important idea that emerged during the process of writing the standard was that the principal sound for a program should take up only the digital bit space required by that program. The idea was born that programs fall into production categories and may be classified by the utilization of loudspeaker channels [17]. The categories include:

- **1/0—one front center channel, no surround**. 1/0 is most likely to be used in programs such as news, which have exceedingly strict production time requirements. The advantage in having a distinct monaural mode is that those end users having a center channel loudspeaker will hear the presentation over only that one loudspeaker, with an attendant improvement over hearing mono presented over two loudspeakers.

- **2/0—conventional 2-channel stereo**. 2/0 is intended principally for pre-existing 2-channel program material. It is also useful for film production recorded in the Dolby Stereo or Ultra Stereo formats with a 4:2:4 amplitude-phase matrix (for which there is an accompanying indicator flag to toggle surround decoding on at the receiver).

- **3/0—left, center, and right front channels**. 3/0 was expected to be used for programs in which stereo is useful but surround sound effects are not, such as an interview program with a panel of experts.

- **3/2/.1—left, center, right front, left and right surround, and a low-frequency effects channel**. 3/2/.1 was expected to be used primarily for films and entertainment programming, matching the current motion picture production practice.

Monitoring

Aural monitoring of program production is a critical element in the production chain [17]. Although the monitor system—with its equalizers, power amplifiers, and loudspeakers—is not in the signal path, monitoring under highly standardized conditions has helped the film industry to make an extremely interchangeable product for many years. With strict monitor standards, there is less variation in program production, and the differences that remain are the result of the director's creative intent. Monitor standards must address the following criteria:

- Room acoustics for monitor spaces

- Physical loudspeaker arrangement

- Loudspeaker and electronics requirements

- Electroacoustics performance

Such attention to detail invariably improves the viewing experience.

1.8   References

1. Clement, Pierre, and Eric Gourmelen: "Internet and Television Convergence: IP and MPEG-2 Implementation Issues," *Proceedings of the 33rd SMPTE Advanced Motion Imaging Conference*, SMPTE, White Plains, N.Y., February 1999.

2. Erez, Beth: "Protecting content in the Digital Home," *Proceedings of the 33rd SMPTE Advanced Motion Imaging Conference*, SMPTE, White Plains, N.Y., pp. 231–238, February 1999.

3. Thomas, Gomer: "ATSC Datacasting—Opportunities and Challenges," *Proceedings of the 33rd SMPTE Advanced Motion Imaging Conference*, SMPTE, White Plains, N.Y., pp. 307–314, February 1999.

4. Robin, Michael: "Digital Resolution," *Broadcast Engineering*, Intertec Publishing, Overland Park, Kan., pp. 44–48, April 1998.

5. Venkat, Giri: "Understanding ATSC Datacasting—A Driver for Digital Television," *Proceedings of the NAB Broadcast Engineering Conference*, National Association of Broadcasters, Washington, D.C., pp. 113–116, 1999.

6. Hoffman, Gary A.: "IEEE 1394: The A/V Digital Interface of Choice," 1394 Technology Association Technical Brief, 1394 Technology Association, Santa Clara, Calif., 1999.

7. "Advanced Television Enhancement Forum Specification," Draft, Version 1.1r26 updated 2/2/99, ATVEF, Portland, Ore., 1999.

8. *NAB TV TechCheck*: National Association of Broadcasters, Washington, D.C., February 1, 1999.

9. Wugofski, T. W.: "A Presentation Engine for Broadcast Digital Television*," International Broadcasting Convention Proceedings*, IBC, London, England, pp. 451–456, 1999.

10. *NAB TV TechCheck*: "CEA Establishes Definitions for Digital Television Products," National Association of Broadcasters, Washington, D.C., September 1, 2000.

11. *NAB TV TechCheck:* "FCC Adopts Rules for Labeling of DTV Receivers," National Association of Broadcasters, Washington, D.C., September 25, 2000.

12. Thorpe, Laurence J.: "Applying High-Definition Television," *Television Engineering Handbook*, rev. ed., K. B. Benson and Jerry C. Whitaker (eds.), McGraw-Hill, New York, p. 23.4, 1991.

13. Benson, K. B., and D. G. Fink: *HDTV: Advanced Television for the 1990s*, McGraw-Hill, New York, 1990.

14. SMPTE Recommended Practice RP 199-1999, "Mapping of Pictures in Wide-Screen (16:9) Scanning Structure to Retain Original Aspect Ratio of the Work," SMPTE, White Plains, N.Y., 1999.

15. Torick, Emil L.: "HDTV: High Definition Video—Low Definition Audio?," *1991 HDTV World Conference Proceedings*, National Association of Broadcasters, Washington, D.C., April 1991.

16. Holman, Tomlinson: "The Impact of Multi-Channel Sound on Conversion to ATV," *Perspectives on Wide Screen and HDTV Production*, National Association of Broadcasters, Washington, D.C., 1995.

17. Holman, Tomlinson: "Psychoacoustics of Multi-Channel Sound Systems for Television," *Proceedings of HDTV World*, National Association of Broadcasters, Washington, D.C., 1992.

1.9   Bibliography

Baldwin, M. Jr.: "The Subjective Sharpness of Simulated Television Images," *Proceedings of the IRE*, vol. 28, July 1940.

Belton, J.: "The Development of the CinemaScope by Twentieth Century Fox," *SMPTE Journal*, vol. 97, SMPTE, White Plains, N.Y., September 1988.

Fink, D. G.: "Perspectives on Television: The Role Played by the Two NTSCs in Preparing Television Service for the American Public," *Proceedings of the IEEE*, vol. 64, IEEE, New York, September 1976.

Fink, D. G: *Color Television Standards*, McGraw-Hill, New York, 1986.

Fink, D. G, et. al.: "The Future of High Definition Television," *SMPTE Journal*, vol. 9, SMPTE, White Plains, N.Y., February/March 1980.

Fujio, T., J. Ishida, T. Komoto, and T. Nishizawa: "High-Definition Television Systems—Signal Standards and Transmission," *SMPTE Journal*, vol. 89, SMPTE, White Plains, N.Y., August 1980.

Hamasaki, Kimio: "How to Handle Sound with Large Screen," *Proceedings of the ITS*, International Television Symposium, Montreux, Switzerland, 1991.

Hubel, David H.: *Eye, Brain and Vision*, Scientific American Library, New York, 1988.

Judd, D. B.: "The 1931 C.I.E. Standard Observer and Coordinate System for Colorimetry," *Journal of the Optical Society of America*, vol. 23, 1933.

Keller, Thomas B.: "Proposal for Advanced HDTV Audio," *1991 HDTV World Conference Proceedings*, National Association of Broadcasters, Washington, D.C., April 1991.

Kelly, R. D., A. V. Bedbord, and M. Trainer: "Scanning Sequence and Repetition of Television Images," *Proceedings of the IRE*, vol. 24, April 1936.

Kelly, K. L.: "Color Designation of Lights," *Journal of the Optical Society of America*, vol. 33, 1943.

Lagadec, Roger, Ph.D.: "Audio for Television: Digital Sound in Production and Transmission," *Proceedings of the ITS*, International Television Symposium, Montreux, Switzerland, 1991.

Miller, Howard: "Options in Advanced Television Broadcasting in North America," *Proceedings of the ITS*, International Television Symposium, Montreux, Switzerland, 1991.

Pitts, K. and N. Hurst: "How Much Do People Prefer Widescreen (16×9) to Standard NTSC (4×3)?," *IEEE Transactions on Consumer Electronics*, IEEE, New York, August 1989.

Pointer, R. M.: "The Gamut of Real Surface Colors, *Color Res. App.*, vol. 5, 1945.

Slamin, Brendan: "Sound for High Definition Television," *Proceedings of the ITS*, International Television Symposium, Montreux, Switzerland, 1991.

Suitable Sound Systems to Accompany High-Definition and Enhanced Television Systems: Report 1072. Recommendations and Reports to the CCIR, 1986. Broadcast Service—Sound. International Telecommunications Union, Geneva, 1986.

2

Video and Audio Compression

Jerry C. Whitaker, Editor

2.1 Introduction

Virtually all applications of video and visual communication deal with an enormous amount of data. Because of this, compression is an integral part of most modern digital video applications. In fact, compression is essential to the ATSC DTV system, and therefore to interactive television [1].

A number of existing and proposed video-compression systems employ a combination of processing techniques. Any scheme that becomes widely adopted can enjoy economies of scale and reduced market confusion. Timing, however, is critical to market acceptance of any standard. If a standard is selected well ahead of market demand, more cost-effective or higher-performance approaches may become available before the market takes off. On the other hand, a standard may be merely academic if it is established after alternative schemes already have become well entrenched in the marketplace.

These forces are shaping the video technology of the future. Any number of scenarios have been postulated as to the hardware and software that will drive the interactive services of the future. One thing is certain, however: It will revolve around compressed video and audio signals.

2.2 Transform Coding

In technical literature, countless versions of different coding techniques can be found [2]. Despite the large number of techniques available, one that comes up regularly (in a variety of flavors) during discussions about transmission standards is *transform coding* (TC).

Transform coding is a universal bit-rate-reduction method that is well suited for both large and small bit rates. Furthermore, because of several possibilities that TC offers for exploiting the visual inadequacies of the human eye, the subjective impression given by the resulting picture is frequently better than with other methods. If the intended bit rate turns out to be insufficient, the effect is seen as a lack of sharpness, which is less disturbing (subjectively) than coding errors such as frayed edges or noise with a structure. Only at very low bit rates does TC produce a particularly noticeable artifact: the *blocking effect.*

Because all pictures do not have the same statistical characteristics, the optimum transform is not constant, but depends on the momentary picture content that has to be coded. It is possible, for example, to recalculate the optimum transform matrix for every new frame to be transmitted, as is performed in the *Karhunen-Loeve transform* (KLT). Although the KLT is efficient in terms of ultimate performance, it is not typically used in practice because investigating each new picture to find the best transform matrix is usually too demanding. Furthermore, the matrix must be indicated to the receiver for each frame, because it must be used in decoding of the relevant inverse transform. A practical compromise is the *discrete cosine transform* (DCT). This transform matrix is constant and is suitable for a variety of images; it is sometimes referred to as "quick KLT."

The DCT is a near relative of the *discrete Fourier transform* (DFT), which is widely used in signal analysis. Similar to DFT techniques, DCT offers a reliable algorithm for quick execution of matrix multiplication.

The main advantage of DCT is that it *decorrelates* the pixels efficiently; put another way, it efficiently converts statistically dependent pixel values into independent coefficients. In so doing, DCT packs the signal energy of the image block onto a small number of coefficients. Another significant advantage of DCT is that it makes available a number of fast implementations. A block diagram of a DCT-based coder is shown in Figure 2.1.

In addition to DCT, other transforms are practical for data compression, such as the *Slant transform* and the *Hadamard transform* [3].

2.2.1 Planar Transform

The similarities of neighboring pixels in a video image are not only line- or column-oriented, but also area-oriented [2]. To make use of these *neighborhood relationships*, it is desirable to transform not only in lines and columns, but also in areas. This can be achieved by a *planar transform*. In practice,

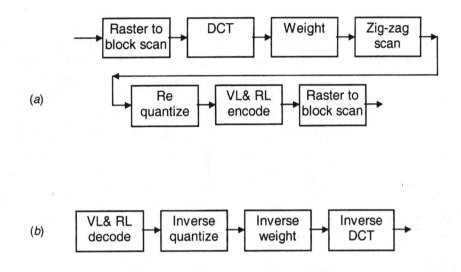

Figure 2.1 Block diagram of a sequential DCT codec: (*a*) encoder, (*b*) decoder. (*From* [2]. *Used with permission.*)

separable transforms are used almost exclusively. A separable planar transform is nothing more than the repeated application of a simple transform. It is almost always applied to square picture segments of size $N \times N$, and it progresses in two steps, as illustrated in Figure 2.2. First, all lines of the picture segments are transformed in succession, then all rows of the segments calculated in the first step are transformed.

In textbooks, the planar transform frequently is called a *2D transform*. The transform is, in principle, possible for any segment forms—not just square ones [4]. Consequently, for a segment of the size $N \times N$, $2N$ transforms are used. The coefficients now are no longer arranged as vectors, but as a matrix. The coefficients of the i lines and the j columns are called c_{ij} (i, $j = 1 \ldots N$). Each of these coefficients no longer represents a basic vector, but a *basic picture*. In this way, each $N \times N$ picture segment is composed of $N \times N$ different basic pictures, in which each coefficient gives the weighting of a particular basic picture. Figure 2.3 shows the basic pictures of the coefficients c_{11} and c_{23} for a planar 4×4 DCT. Because c_{11} represents the dc part, it is called the *dc coefficient*; the others are appropriately called the *ac coefficients*.

The planar transform of television pictures in the interlaced format is somewhat problematic. In moving regions of the picture, depending on the

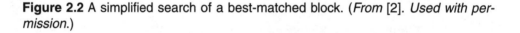

110	106	98	92
112	105	97	90
109	107	97	94
111	107	95	91

⇒

203	14	−1	−0.4
202	16.5	0	0.7
203.5	12.5	−0.5	−2.5
202	16.3	0	−2.4

⇒

405	29.7	−0.8	−2.3
0.3	−0.4	−0.5	2.2
−0.3	0.7	−0.3	−0.5
1.3	−3.2	−0.6	−1.6

4x4 picture segment...   ...after DCT of lines...   ...after DCT of rows

Figure 2.2 A simplified search of a best-matched block. (*From* [2]. *Used with permission.*)

speed of motion, the similarities of vertically neighboring pixels of a frame are lost because changes have occurred between samplings of the two halves of the picture. Consequently, interlaced scanning may cause the performance of the system (or *output concentration*) to be greatly weakened, compared with progressive scanning. Well-tuned algorithms, therefore, try to detect stronger movements and switch to a transform in one picture half (i.e., field) for these picture regions [5]. However, the coding in one-half of the picture is less efficient because the correlation of vertically neighboring pixels is weaker than in the full picture of a static scene. Simply stated, if the picture sequences are interlaced, the picture quality may be influenced by the motion content of the scene to be coded.

2.2.2 Interframe Transform Coding

With common algorithms, compression factors of approximately 8 can be achieved while maintaining good picture quality [2]. To achieve higher factors, the similarities between successive frames must be exploited. The nearest approach to this goal is the extension of the DCT in the time dimension. A drawback of such *cubic* transforms is the increase in calculation effort, but the greatest disadvantage is the higher memory requirement: for an $8 \times 8 \times 8$ DCT, at least seven frame memories would be needed. Much simpler is the

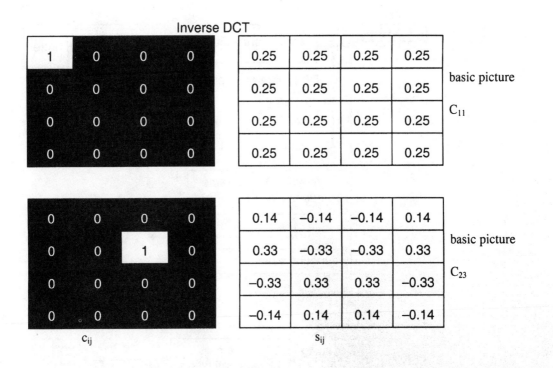

Inverse DCT

1	0	0	0
0	0	0	0
0	0	0	0
0	0	0	0

0.25	0.25	0.25	0.25
0.25	0.25	0.25	0.25
0.25	0.25	0.25	0.25
0.25	0.25	0.25	0.25

basic picture

C_{11}

0	0	0	0
0	0	1	0
0	0	0	0
0	0	0	0

c_{ij}

0.14	−0.14	−0.14	0.14
0.33	−0.33	−0.33	0.33
−0.33	0.33	0.33	−0.33
−0.14	0.14	0.14	−0.14

s_{ij}

basic picture

C_{23}

Figure 2.3 The mechanics of motion-compensated prediction. Shown are the pictures for a planar 4×4 DCT. Element C_{11} is located at row 1, column 1; element C_{23} is located at row 2, column 3. Note that picture C_{11} values are constant, referred to as dc coefficients. The changing values shown in picture C_{23} are known as ac coefficients. (*From* [2]. *Used with permission.*)

hybrid DCT, which also efficiently codes pictures with moving objects. This method comprises, almost exclusively, a motion-compensated *difference pulse-code-modulation* (DPCM) technique; instead of each picture being transferred individually, the motion-compensated difference of two successive frames is coded.

DPCM is, in essence, predictive coding of sample differences. DPCM can be applied for both *interframe coding*, which exploits the temporal redundancy of the input image, and *intraframe coding*, which exploits the spatial redundancy of the image. In the intraframe mode, the difference is calculated using the values of two neighboring pixels of the same frame. In the interframe mode, the difference is calculated using the value of the same pixel on two consecutive frames. In either mode of operation, the value of the target pixel is predicted using the reconstructed values of the previously

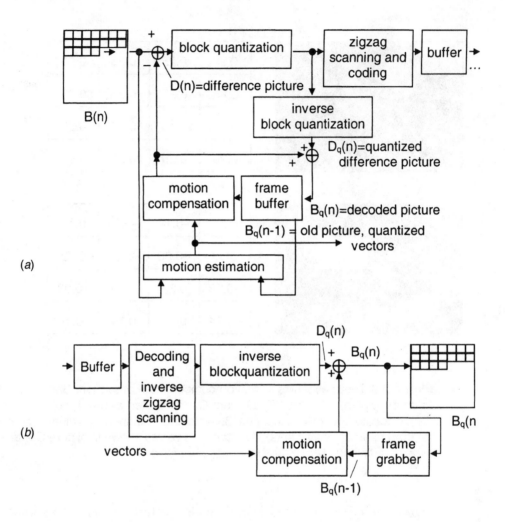

Figure 2.4 Overall block diagram of a DPCM system: (*a*) encoder, (*b*) decoder. (*From* [2]. *Used with permission.*)

coded neighboring pixels. This value is then subtracted from the original value to form the differential image value. The differential image is then quantized and encoded. Figure 2.4 illustrates an end-to-end DPCM system.

2.3 The JPEG Standard

The JPEG (Joint Photographic Experts Group) standard is enjoying commercial use today in a wide variety of applications. Because JPEG is the product of a committee, it is not surprising that it includes more than one fixed encoding/decoding scheme. It can be thought of as a family of related compression techniques from which designers can choose, based upon suitability for the application under consideration. The four primary JPEG family members are [2]:

- Sequential DCT-based
- Progressive DCT-based
- Sequential lossless
- Hierarchical

As JPEG has been adapted to other environments, additional JPEG schemes have come into practice. JPEG is designed for still images and offers reduction ratios of 10:1 to 50:1. The algorithm is symmetrical—the time required for encoding and decoding is essentially the same. There is no need for motion compensation, and there are no provisions for audio in the basic standard.

The JPEG specification, like MPEG-1 and MPEG-2, often is described as a "tool kit" of compression techniques. Before looking at specifics, it will be useful to examine some of the basics.

2.3.1 Compression Techniques

As discussed briefly in previous sections, a compression system reduces the volume of data by exploiting spatial and temporal redundancies and by eliminating the data that cannot be displayed suitably by the associated display or imaging devices. The main objective of compression is to retain as little data as possible, just sufficient to reproduce the original images without causing unacceptable distortion of the images [1]. A compression system consists of the following components:

- *Digitization, sampling, and segmentation*: Steps that convert analog signals on a specified grid of picture elements into digital representations and then divide the video input—first into frames, then into blocks.

- *Redundancy reduction*: The decorrelation of data into fewer useful data bits using certain invertible transformation techniques.

- *Entropy reduction*: The representation of digital data using fewer bits by dropping less significant information. This component causes distortion; it is the main contributor in *lossy* compression.

- *Entropy coding*: The assignment of code words (bit strings) of shorter length to more likely image symbols and code words of longer length to less likely symbols. This minimizes the average number of bits needed to code an image.

Key terms important to the understanding of this topic include the following:

- *Motion compensation*: The coding of video segments with consideration to their displacements in successive frames.

- *Spatial correlation*: The correlation of elements within a still image or a video frame for the purpose of bit-rate reduction.

- *Spectral correlation*: The correlation of different color components of image elements for the purpose of bit-rate reduction.

- *Temporal correlation*: The correlation between successive frames of a video file for the purpose of bit-rate reduction.

- *Quantization compression*: The dropping of the less significant bits of image values to achieve higher compression.

- *Intraframe coding*: The encoding of a video frame by exploiting spatial redundancy within the frame.

- *Interframe coding*: The encoding of a frame by predicting its elements from elements of the previous frame.

The removal of spatial and temporal redundancies that exist in natural video imagery is essentially a lossless process. Given the correct techniques, an exact replica of the image can be reproduced at the viewing end of the system. Such lossless techniques are important for medical imaging applications and other demanding uses. These methods, however, may realize only low compression efficiency (on the order of approximately 2:1). For video, a much higher compression ratio is required. Exploiting the inherent limitations of the *human visual system* (HVS) can result in compression ratios of 50:1 or higher [6]. These limitations include the following:

- Limited luminance response and very limited color response

- Reduced sensitivity to noise in high frequencies, such as at the edges of objects

- Reduced sensitivity to noise in brighter areas of the image

The goal of compression, then, is to discard all information in the image that is not absolutely necessary from the standpoint of what the HVS is capable of resolving. Such a system can be described as *psychovisually lossless*.

2.3.2 DCT and JPEG

DCT is one of the building blocks of the JPEG standard. All JPEG DCT-based coders start by portioning the input image into nonoverlapping blocks of 8×8 picture elements. The 8-bit samples are then level-shifted so that the values range from -128 to $+127$. A fast Fourier transform then is applied to shift the elements into the frequency domain. Huffman coding is mandatory in a baseline system; other arithmetic techniques can be used for entropy coding in other JPEG modes. The JPEG specification is independent of color or gray scale. A color image typically is encoded and decoded in the *YUV* color space with four pixels of *Y* for each *U, V* pair.

In the *sequential DCT*-based mode, processing components are transmitted or stored as they are calculated in one single pass. Figure 2.5 provides a simplified block diagram of the coding system.

The *progressive DCT*-based mode can be convenient when it takes a perceptibly long time to send and decode the image. With progressive DCT-based coding, the picture first will appear blocky, and the details will subsequently appear. A viewer may linger on an interesting picture and watch the details come into view or move onto something else, making this scheme well suited, for example, to the Internet.

In the *lossless* mode, the decoder reproduces an exact copy of the digitized input image. The compression ratio, naturally, varies with picture content. The varying compression ratio is not a problem for sending still photos, but presents significant challenges for sequential images that must be viewed in real time.

The efficiency of JPEG coding for still images led to the development of *motion* JPEG (M-JPEG) for video applications, primarily studio use. Motion JPEG uses intraframe compression, where each frame is treated as an individual signal; a series of frames is basically a stream of JPEG signals. The benefit of this construction is easy editing, making the technique a good choice for nonlinear editing applications. Also, any individual frame is self-

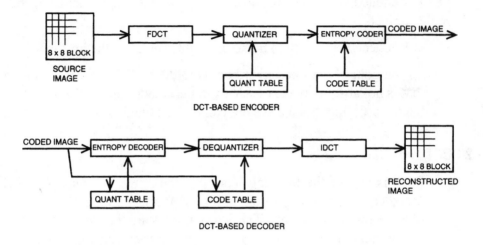

Figure 2.5 Block diagram of a DCT-based image-compression system. Note how the 8 × 8 source image is processed through a *forward-DCT* (FDCT) encoder and related systems to the *inverse-DCT* (IDCT) decoder and reconstructed into an 8 × 8 image. (*From* [1]. *Used with permission.*)

supporting and can be accessed as a stand-alone image. The intraframe system is based, again, on DCT. Because a picture with high-frequency detail will generate more data than a picture with low detail, the data stream will vary. This is problematic for most real-time systems, which would prefer to see a constant data rate at the expense of varying levels of quality. The symmetry in complexity of decoders and encoders is another consideration in this regard.

The major disadvantage of motion JPEG is bandwidth and storage requirements. Because stand-alone frames are coded, there is no opportunity to code only the differences between frames (to remove redundancies).

M-JPEG, in its basic form, addresses only the video—not the audio—component. Many of the early problems experienced by users concerning portability of M-JPEG streams stemmed from the methods used to include audio in the data stream. Because the location of the audio may vary from one unit to the next, some decoder problems were experienced [7].

Table 2.1 Participants in Early MPEG Proceedings (*After* [2].)

Computer Manufacturers	IC Manufacturers
Apple	Brooktree
DEC	C-Cube
Hewlett-Packard	Cypress
IBM	Inmos
NEC	Intel
Olivetti	IIT
Sun	LSI Logic
	Motorola
Software Suppliers	National Semiconductor
Microsoft	Rockwell
Fluent Machines	SGS-Thomson
Prism	Texas Instruments
	Zoran
Audio/Visual Equipment Manufacturers	
Dolby	**Universities/Research**
JVC	Columbia University
Matsushita	Massachusetts Institute of Technology
Philips	DLR
Sony	University of Berlin
Thomson Consumer Electronics	Fraunhofer Gesellschaft
	University of Hannover

2.4 The MPEG Standard

The Moving Picture Experts Group (MPEG) was founded in 1988 with the objective of specifying an audio/visual decompression system, composed of three basic elements, which the sponsoring organization (the *International Standards Organization,* or ISO) calls "parts." They are as follows:

- *Part 1—Systems*: Describes the audio/video synchronization, multiplexing, and other system-related elements

- *Part 2—Video*: Contains the coded representation of video data and the decoding process

- *Part 3—Audio*: Contains the coded representation of audio data and the decoding process

The basic MPEG system, finalized in 1992, was designated MPEG-1. Shortly thereafter, work began on MPEG-2. The first three stages (systems, video, and audio) of the MPEG-2 standard were agreed to in November 1992. Table 2.1 lists the companies and organizations participating in the early MPEG work. Because of their combined efforts, the MPEG standards have achieved broad market acceptance.

As might be expected, the techniques of MPEG-1 and MPEG-2 are similar, and their syntax is rather extensible.

2.4.1 Basic Provisions

When trying to settle on a specification, it is always important to have a target application in mind [1]. The definition of MPEG-1 (also known as ISO/IEC 11172) was driven by the desire to encode audio and video onto a compact disc. A CD is defined to have a constant bit rate of 1.5 Mbits/s. With this constrained bandwidth, the target video specifications were:

- Horizontal resolution of 360 pixels

- Vertical resolution of 240 for NTSC, and 288 for PAL and SECAM

- Frame rate of 30 Hz for NTSC, 25 for PAL and SECAM, and 24 for film

A detailed block diagram of an MPEG-1 codec (coder-decoder) is shown in Figure 2.6.

MPEG uses the JPEG standard for intraframe coding by first dividing each frame of the image into 8 × 8 blocks, then compressing each block independently using DCT-based techniques. Interframe coding is based on *motion compensation* (MC) prediction that allows bidirectional temporal prediction. A block-matching algorithm is used to find the best-matched block, which may belong to either the past frame (*forward prediction*) or the future frame (*backward prediction*). The best-matched block may—in fact—be the average of two blocks, one from the previous and the other from the next frame of the target frame (*interpolation*). In any case, the placement of the best-matched block(s) is used to determine the motion vector(s); blocks predicted on the basis of interpolation have two motion vectors. Frames that are bidirectionally predicted never are used themselves as reference frames.

2.4.2 Motion Compensation

At this point, it is appropriate to take a closer look at MC prediction [1]. For motion-compensated interframe coding, the target frame is divided into non-overlapping fixed-size blocks, and each block is compared with blocks of the same size in some reference frame to find the best match. To limit the search, a small *neighborhood* is selected in the reference frame, and the search is performed by *stepwise translation* of the target block.

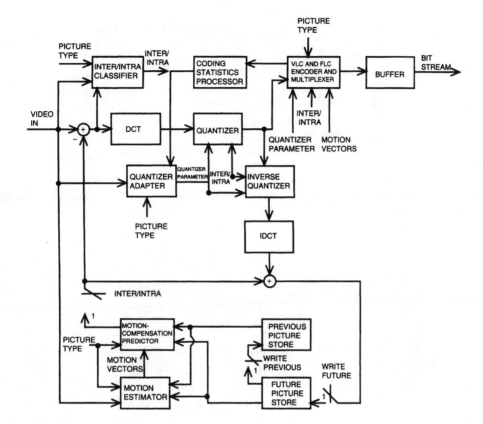

Figure 2.6 A typical MPEG-1 codec: (*a*) encoder, (*b*) decoder. (*After* [8].)

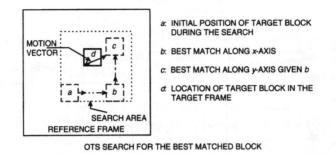

OTS SEARCH FOR THE BEST MATCHED BLOCK

Figure 2.7 A simplified search of a best-matched block. (*From* [1]. *Used with permission.*)

To reduce mathematical complexity, a simple block-matching criterion, such as the mean of the absolute difference of pixels, is used to find a best-matched block. The position of the best-matched block determines the displacement of the target block, and its location is denoted by a (motion) vector.

Block matching is computationally expensive; therefore, a number of variations on the basic theme have been developed. A simple method known as *OTS* (one-at-a-time search) is shown in Figure 2.7. First, the target block is moved along in one direction and the best match found, then it is moved along perpendicularly to find the best match in that direction. Figure 2.7 portrays the target frame in terms of the best-matched blocks in the reference frame.

2.4.3 Putting It All Together

MPEG is a standard built upon many elements [1]. Figure 2.8 shows a *group of pictures* (GOP) of 14 frames with two different orderings. Pictures marked *I* are intraframe-coded. A *P*-picture is predicted using the most recently encoded *P*- or *I*-picture in the sequence. A macroblock in a *P*-picture can be coded using either the intraframe or the forward-predicted method. A *B*-picture macroblock can be predicted using either or both of the previous or the next *I*- and/or *P*-pictures. To meet this requirement, the transmission order and display order of frames are different. The two orders also are shown in Figure 2.8.

The MPEG-coded bit stream is divided into several layers, listed in Table 2.2. The three primary layers are:

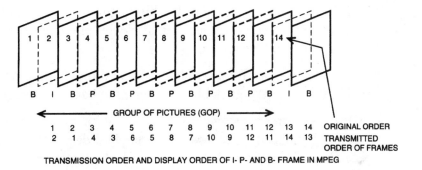

Figure 2.8 Illustration of *I*-frames, *P*-frames, and *B*-frames. (*From* [1]. *Used with permission.*)

Table 2.2 Layers of the MPEG-2 Video Bit-Stream Syntax (*After* [2].)

Syntax layer	Functionality
Video sequence layer	Context unit
Group of pictures (GOP) layer	Random access unit: video coding
Picture layer	Primary coding unit
Slice layer	Resynchronization unit
Macroblock layer	Motion-compensation unit
Block layer	DCT unit

- *Video sequence*, the outermost layer, which contains basic global information such as the size of frames, bit rate, and frame rate.

- *GOP layer*, which contains information on fast search and random access of the video data. The length of a GOP is arbitrary.

- *Picture layer*, which contains a coded frame. Its header defines the type (*I*, *P*, *B*) and the position of the frame in the GOP.

Several of the major differences between MPEG and other compression schemes (such as JPEG) include the following:

- MPEG focuses on video. The basic format uses a single color space (Y, C_r, C_b), a limited range of resolutions and compression ratios, and has built-in mechanisms for handling audio.

- MPEG takes advantage of the high degree of commonality between pictures in a video stream and the typically predictable nature of movement (*inter-picture encoding*).

- MPEG provides for a constant bit rate through adjustable variables, making the format predictable with regard to bandwidth requirements.

MPEG specifies the syntax for storing and transmitting compressed data and defines the decoding process. The standard does not, however, specify how encoding should be performed. Such implementation considerations are left to the manufacturers of encoding systems. Still, all conforming encoders must produce valid MPEG bit streams that can be decompressed by any MPEG decoder. This approach is, in fact, one of the strengths of the MPEG standard; because encoders are allowed to use proprietary but compliant algorithms, a variety of implementations is possible and, indeed, encouraged.

As mentioned previously, MPEG is actually a collection of standards, each suited to a particular application or group of applications, including:

- **MPEG-1**, the original implementation, targeted at multimedia uses. The MPEG-1 algorithm is intended basically for compact disc bit rates of approximately 1.5–2.0 Mbits/s. MPEG-1 supports 525- and 625-type signal structures in progressive form with 204/288 lines per frame, sequential-scan frame rates of 29.97 and 25 per second, and 352 pixels per line. The coding of high-motion signals does not produce particularly good results, however. As might be expected, as the bit rate is reduced (compression increased), the output video quality gradually declines. The overall bit-rate reduction ratios achievable are about 6:1 with a bit rate of 6 Mbits/s and 200:1 at 1.5 Mbits/s. The MPEG-1 system is not symmetrical; the compression side is more complex and expensive than the decompression process, making the system ideal for broadcast-type applications in which there are far more decoders than encoders.

- **MPEG-2**, which offers full ITU-R Rec. 601 resolution for professional and broadcast uses, and is the chosen standard for the ATSC DTV system and the European DVB suite of applications.

- **MPEG-3**, originally targeted at high-definition imaging applications. Subsequent to development of the standard, however, key specifications of MPEG-3 were absorbed into MPEG-2. Thus, MPEG-3 is no longer in use.

- **MPEG-4,** a standard that uses very low bit rates for teleconferencing and related applications requiring high bit efficiency. Like MPEG-2, MPEG-4 is a collection of tools that can be grouped into profiles and levels for different video applications. The MPEG-4 video coding structure ranges from a *very low bit rate video* (VLBV) level, which includes algorithms and tools for data rates between 5 kbits/s and 64 kbits/s, to ITU-R. Rec. 601 quality video at 2 Mbits/s. MPEG-4 does not concern itself directly with the error protection required for specific channels, such as cellular ratio, but it has made improvements in the way payload bits are arranged so that recovery is more robust.

- **MPEG-7,** not really a compression scheme at all, but rather a "multimedia content description interface." MPEG-7 is an attempt to provide a standard means of describing multimedia content.

2.4.4 Profiles and Levels

Six *profiles* and four *levels* describe the organization of the basic MPEG-2 standard. A *profile* is a subset of the MPEG-2 bit-stream syntax with restrictions on the parts of the MPEG algorithm used. Profiles are analogous to features, describing the available characteristics. A *level* constrains general parameters such as image size, data rate, and decoder buffer size. Levels describe, in essence, the upper bounds for a given feature and are analogous to performance specifications.

By far the most popular element of the MPEG-2 standard for professional video applications is the *Main Profile* in conjunction with the *Main Level* (described in the jargon of MPEG as Main Profile/Main Level), which gives an image size of 720 × 576, a data rate of 15 Mbits/s, and a frame rate of 30 frames/s. All higher profiles are capable of decoding Main Profile/Main Level streams.

Table 2.3 lists the basic MPEG-2 classifications. With regard to the table, the following generalizations can be made:

- The three key flavors of MPEG-2 are Main Profile/Low Level (source input format, or SIF), Main Profile/Main Level (Main), and Studio Profile/Main Level (Studio).

- The SIF Main Profile/Low Level offers the best picture quality for bit rates below about 5 Mbits/s. This provides generally acceptable quality for interactive and multimedia applications. The SIF profile has replaced MPEG-1 in some applications.

Table 2.3 Common MPEG Profiles and Levels in Simplified Form (*After* [2] *and* [12].)

Profile	General Specifications	Parameter	Low	Main (ITU 601)	High 1440 (HD, 4:3)	High (HD, 16:9)
Simple	Pictures: I, P Chroma: 4:2:0	Image size[1]		720×576		
		Image frequency[2]		30		
		Bit rate[3]		15		
Main	Pictures: I, P, B Chroma: 4:2:0	Image size	325×288	720×576	1440×1152	1920×1152
		Image frequency	30	30	60	60
		Bit rate	4	15	100	80
SNR-Scalable	Pictures: I, P, B Chroma: 4:2:0	Image size	325×288	720×576		
		Image frequency	30	30		
		Bit rate	3, 4[4]	15		
Spatially-Scalable	Pictures: I, P, B Chroma: 4:2:0	Image size			720×576	
		Image frequency			30	
		Bit rate			15	
	Enhancement Layer[5]	Image size			1440×1152	
		Image frequency			60	
		Bit rate			40, 60[6]	
High[7]	Pictures: I, P, B Chroma: 4:2:2	Image size		720×576	1440×1152	1920×1152
		Image frequency		30	60	60
		Bit rate		20	80	100
Studio	Pictures: I, P, B Chroma: 4:2:2	Image size		720×608		
		Image frequency		30		
		Bit rate		50		

Notes:
[1] Image size specified as samples/line × lines/frame
[2] Image frequency in frames/s
[3] Bit rate in Mbits/s
[4] For *Enhancement Layer 1*
[5] For *Enhancement Layer 1*, except as noted by [6] for *Enhancement Layer 2*
[7] For simplicity, *Enhancement Layers* not specified individually

- The Main Profile/Main Level grade offers the best picture quality for conventional video systems at rates from about 5 to 15 Mbits/s. This provides good quality for broadcast applications such as play-to-air, where four generations or fewer typically are required.

- The Studio Profile offers high quality for multiple-generation conventional video applications, such as post-production.

- The High Profile targets HDTV applications.

2.4.5 Studio Profile

Despite the many attributes of MPEG-2, the Main Profile/Main Level remains a less-than-ideal choice for conventional video production because the larger GOP structure makes individual frames hard to access. For this reason, the 4:2:2 *Studio Profile* was developed. The Studio Profile expands upon the 4:2:0 sampling scheme of MPEG-1 and MPEG-2. In essence, "standard MPEG" samples the full luminance signal, but ignores half of the chrominance information, specifically the color coordinate on one axis of the color grid. Studio Profile MPEG increases the chrominance sampling to 4:2:2, thereby accounting for both axes on the color grid by sampling every other element. This enhancement provides better replication of the original signal.

The Studio Profile is intended principally for editing applications, where multiple iterations of a given video signal are required or where the signal will be compressed, decompressed, and recompressed several times before it is finally transmitted or otherwise finally displayed.

SMPTE 308M

SMPTE standard 308M is intended for use in high-definition television production, contribution, and distribution applications [9]. It defines bitstreams, including their syntax and semantics, together with the requirements for a compliant decoder for 4:2:2 Studio Profile at High Level. As with the other MPEG standards, 308M does not specify any particular encoder operating parameters.

The MPEG-2 4:2:2 Studio Profile is defined in ISO/IEC 13818-2, and in SMPTE 308M, only those additional parameters necessary to define the 4:2:2 Studio Profile at High Level are specified. The primary differences are: 1) the upper bounds for sampling density are increased to 1920 samples/line, 1088 lines/frame, and 60 frames/s; 2) the upper bounds for the luminance sample rate is set at 62,668,800 samples/s; and 3) the upper bounds for bit rates is set at 300 Mbits/s.

2.5   MPEG-2 Features of Importance for DTV

The primary application of interest when the MPEG-2 standard was first defined was "true" television broadcast resolution, as specified by ITU-R Rec. 601. This is roughly four times more picture information than the MPEG-1 standard provides. MPEG-2 is a superset, or extension, of MPEG-1. As such, an MPEG-2 decoder also should be able to decode an MPEG-1 stream. This broadcast version adds to the MPEG-1 toolbox provisions for dealing with interlace, graceful degradation, and hierarchical coding.

Although MPEG-1 and MPEG-2 each were specified with a particular range of applications and resolutions in mind, the committee's specifications form a set of techniques that support multiple coding options, including picture types and macroblock types. Many variations exist with regard to picture size and bit rates. Also, although MPEG-1 can run at high bit rates and at full ITU-R Rec. 601 resolution, it processes frames, not fields. This fact limits the attainable quality, even at data rates approaching 5 Mbits/s.

The MPEG specifications apply only to decoding, not encoding. The ramifications of this approach are:

- Owners of existing decoding software can benefit from future break-throughs in encoding processing. Furthermore, the suppliers of encoding equipment can differentiate their products by cost, features, encoding quality, and other factors.

- Different schemes can be used in different situations. For example, although *Monday Night Football* must be encoded in real time, a film can be encoded in non-real time, allowing for fine-tuning of the parameters via computer or even a human operator.

2.5.1 MPEG-2 Layer Structure

To allow for a simple yet upgradable system, MPEG-2 defines only the functional elements—syntax and semantics—of coded streams. Using the same system of *I*-, *P*-, and *B*-frames developed for MPEG-1, MPEG-2 employs a 6-layer hierarchical structure that breaks the data into simplified units of information, as listed in Table 2.2.

The top *sequence layer* defines the decoder constraints by specifying the context of the video sequence. The sequence-layer data header contains information on picture format and application-specific details. The second level allows for random access to the decoding process by having a periodic series of pictures; it is fundamentally this GOP layer that provides the bidirectional frame prediction. Intraframe-coded (*I*) frames are the entry-point frames, which require no data from other frames in order to reconstruct. Between the *I*-frames lie the predictive (*P*) frames, which are derived from analyzing previous frames and performing motion estimation. These *P*-frames require about one-third as many bits per frame as *I*-frames. *B*-frames, which lie between two *I*-frames or *P*-frames, are bidirectionally encoded, making use of past and future frames. The *B*-frames require only about one-ninth of the data per frame, compared with *I*-frames.

These different compression ratios for the frames lead to different data rates, so that buffers are required at both the encoder output and the decoder input to ensure that the sustained data rate is constant. One difference between MPEG-1 and MPEG-2 is that MPEG-2 allows for a variety of data-buffer sizes, to accommodate different picture dimensions and to prevent buffer under- and overflows.

The data required to decode a single picture is embedded in the *picture layer*, which consists of a number of horizontal *slice layers*, each containing several macroblocks. Each *macroblock layer*, in turn, is made up of a number of individual blocks. The picture undergoes DCT processing, with the slice layer providing a means of synchronization, holding the precise position of the slice within the image frame.

MPEG-2 places the motion vectors into the coded macroblocks for *P*- and *B*- frames; these are used to improve the reconstruction of predicted pictures. MPEG-2 supports both field- and frame-based prediction, thus accommodating interlaced signals.

The last layer of MPEG-2's video structure is the *block layer*, which provides the DCT coefficients of either the transformed image information for *I*-frames or the residual prediction error of *B*- and *P*- frames.

2.5.2 Slices

Two or more contiguous macroblocks within the same row are grouped together to form *slices* [10]. The order of the macroblocks within a slice is the same as the conventional television raster scan, being from left to right.

Slices provide a convenient mechanism for limiting the propagation of errors. Because the coded bit stream consists mostly of variable-length code words, any uncorrected transmission errors will cause a decoder to lose its sense of code word alignment. Each slice begins with a slice start code. Because the MPEG code word assignment guarantees that no legal combination of code words can emulate a start code, the slice start code can be used to regain the sense of code-word alignment after an error. Therefore, when an error occurs in the data stream, the decoder can skip to the start of the next slice and resume correct decoding.

The number of slices affects the compression efficiency; partitioning the data stream to have more slices provides for better error recovery, but claims bits that could otherwise be used to improve picture quality.

In the DTV system, the initial macroblock of every horizontal row of macroblocks is also the beginning of a slice, with a possibility of several slices across the row.

2.5.3 Pictures, Groups of Pictures, and Sequences

The primary coding unit of a video sequence is the individual video frame or picture [10]. A video picture consists of the collection of slices, constituting the *active picture area*.

A *video sequence* consists of a collection of two or more consecutive pictures. A video sequence commences with a sequence header and is terminated by an end-of-sequence code in the data stream. A video sequence may contain additional sequence headers. Any video-sequence header can serve as an *entry point*. An entry point is a point in the coded video bit stream after which a decoder can become properly initialized and correctly parse the bitstream syntax.

Two or more pictures (frames) in sequence may be combined into a GOP to provide boundaries for interframe picture coding and registration of time code. GOPs are optional within both MPEG-2 and the ATSC DTV system. Figure 2.9 illustrates a typical time sequence of video frames.

I-Frames

Some elements of the compression process exploit only the spatial redundancy within a single picture (frame or field) [10]. These processes constitute intraframe coding, and do not take advantage of the temporal correlation addressed by temporal prediction (interframe) coding. Frames that do not use any interframe coding are referred to as *I*-frames (where "I" denotes *intraframe*-coded). The ATSC video-compression system utilizes both intraframe and interframe coding.

The use of periodic *I*-frames facilitates receiver initializations and channel acquisition (for example, when the receiver is turned on or the channel is changed). The decoder also can take advantage of the intraframe coding

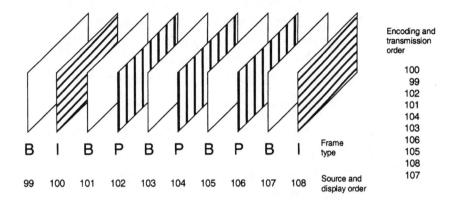

Encoding and
transmission
order

| 100 |
| 99 |
| 102 |
| 101 |
| 104 |
| 103 |
| 106 |
| 105 |
| 108 |
| 107 |

B I B P B P B P B I   Frame type

99 100 101 102 103 104 105 106 107 108   Source and display order

Figure 2.9 Sequence of video frames for the MPEG-2/ATSC DTV system. (*From* [10]. *Used with permission.*)

mode when noncorrectable channel errors occur. With motion-compensated prediction, an initial frame must be available at the decoder to start the prediction loop. Therefore, a mechanism must be built into the system so that if the decoder loses synchronization for any reason, it can rapidly reacquire tracking.

The frequency of occurrence of *I*-pictures may vary and is selected at the encoder. This allows consideration to be given to the need for random access and the location of scene cuts in the video sequence.

P-Frames

P-frames, where the temporal prediction is in the forward direction only, allow the exploitation of interframe coding techniques to improve the overall compression efficiency and picture quality [10]. *P*-frames may include portions that are only intraframe-coded. Each macroblock within a *P*-frame can be either forward-predicted or intraframe-coded.

B-Frames

The *B*-frame is a picture type within the coded video sequence that includes prediction from a future frame as well as from a previous frame [10]. The referenced future or previous frames, sometimes called *anchor frames*, are in all cases either *I*- or *P*-frames.

The basis of the *B*-frame prediction is that a video frame is correlated with frames that occur in the past as well as those that occur in the future. Consequently, if a future frame is available to the decoder, a superior prediction can be formed, thus saving bits and improving performance. Some of the consequences of using future frames in the prediction are:

- The *B*-frame cannot be used for predicting future frames.

- The transmission order of frames is different from the displayed order of frames.

- The encoder and decoder must reorder the video frames, thereby increasing the total latency.

In the example illustrated in Figure 2.9, there is one *B*-frame between each pair of *I*- and *P*-frames. Each frame is labeled with both its display order and transmission order. The *I* and *P* frames are transmitted out of sequence, so the video decoder has both anchor frames decoded and available for prediction.

B-frames are used for increasing the compression efficiency and perceived picture quality when encoding latency is not an important factor. The use of *B*-frames increases coding efficiency for both interlaced- and progressive-scanned material. *B*-frames are included in the DTV system because the increase in compression efficiency is significant, especially with progressive scanning. The choice of the number of bidirectional pictures between any pair of reference (*I* or *P*) frames can be determined at the encoder.

Motion Estimation

The efficiency of the compression algorithm depends on, first, the creation of an estimate of the image being compressed and, second, subtraction of the pixel values of the estimate or prediction from the image to be compressed [10]. If the estimate is good, the subtraction will leave a very small residue to be transmitted. In fact, if the estimate or prediction were perfect, the difference would be zero for all the pixels in the frame of differences, and no new information would need to be sent; this condition can be approached for still images.

If the estimate is not close to zero for some pixels or many pixels, those differences represent information that needs to be transmitted so that the decoder can reconstruct a correct image. The kinds of image sequences that cause large prediction differences include severe motion and/or sharp details.

2.5.4 Vector Search Algorithm

The video-coding system uses motion-compensated prediction as part of the data-compression process [10]. Thus, macroblocks in the current frame of interest are predicted by macroblock-sized regions in previously transmitted frames. Motion compensation refers to the fact that the locations of the macroblock-sized regions in the reference frame can be offset to account for local motions. The macroblock offsets are known as *motion vectors*.

The DTV standard does not specify how encoders should determine motion vectors. One possible approach is to perform an exhaustive search to identify the vertical and horizontal offsets that minimize the total difference between the offset region in the reference frame and the macroblock in the frame to be coded.

2.5.5 Motion-Vector Precision

The estimation of interframe displacement is calculated with half-pixel precision, in both vertical and horizontal dimensions [10]. As a result, the displaced macroblock from the previous frame can be displaced by noninteger displacements and will require interpolation to compute the values of displaced picture elements at locations not in the original array of samples. Estimates for half-pixel locations are computed by averages of adjacent sample values.

2.5.6 Motion-Vector Coding

Motion vectors within a slice are differenced, so that the first value for a motion vector is transmitted directly, and the following sequence of motion-vector differences is sent using *variable-length codes* (VLC) [10]. Motion vectors are constrained so that all pixels from the motion-compensated prediction region in the reference picture fall within the picture boundaries.

2.5.7 Encoder Prediction Loop

The encoder prediction loop, shown in the simplified block diagram of Figure 2.10, is the heart of the video-compression system for DTV [10]. The prediction loop contains a prediction function that estimates the picture values of the next picture to be encoded in the sequence of successive pictures that constitute the TV program. This prediction is based on previous information that is available within the loop, derived from earlier pictures. The

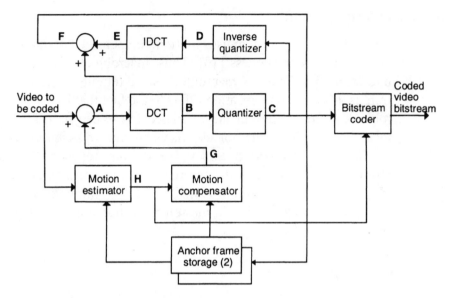

A Pixel-by-pixel prediction errors
B Transformed blocks of prediction errors (DCT coefficients)
C Prediction error DCT coefficients in quantized form
D Quantized prediction error DCT coefficients in standard form
E Pixel-by-pixel prediction errors, degraded by quantization
F Reconstructed pixel values, degraded by quantization
G Motion compensated predicted pixel values
H Motion vectors

Figure 2.10 Simplified encoder prediction loop. (*From* [10]. *Used with permission.*)

transmission of the predicted compressed information works because the same information used to make the prediction also is available at the receiving decoder (barring transmission errors, which are usually infrequent within the primary coverage area).

The subtraction of the predicted picture values from the new picture to be coded is at the core of predictive coding. The goal is to do such a good job of predicting the new values that the result of the subtraction function at the beginning of the prediction loop is zero or close to zero most of the time.

The prediction differences are computed separately for the luminance and two chrominance components before further processing. As explained in previous discussion of *I*-frames, there are times when prediction is not used, for part of a frame or for an entire frame.

Spatial Transform Block—DCT

The image prediction differences (sometimes referred to as *prediction errors*) are organized into 8 × 8 blocks, and a spatial transform is applied to the blocks of difference values [10]. In the intraframe case, the spatial transform is applied to the raw, undifferenced picture data. The luminance and two chrominance components are transformed separately. Because the chrominance data is subsampled vertically and horizontally, each 8 × 8 block of chrominance (C_b or C_r) data corresponds to a 16 × 16 macroblock of luminance data, which is not subsampled.

The spatial transform used is the discrete cosine transform. In principle, applying the IDCT to the transformed array would yield exactly the same array as the original. In that sense, transforming the data does not modify the data, but merely represents it in a different form.

The decoder uses the inverse transformation to approximately reconstruct the arrays that were transformed at the encoder, as part of the process of decoding the received compressed data. The approximation in that reconstruction is controlled in advance during the encoding process for the purpose of minimizing the visual effects of coefficient inaccuracies while reducing the quantity of data that needs to be transmitted.

Quantizer

The process of transforming the original data organizes the information in a way that exposes the spatial frequency components of the images or image differences [10]. Using information about the response of the human visual system to different spatial frequencies, the encoder can selectively adjust the precision of transform coefficient representation. The goal is to include as much information about a particular spatial frequency as necessary—and as possible, given the constraints on data transmission—while not using more precision than is needed, based upon visual perception criteria.

For example, in a portion of a picture that is "busy" with a great deal of detail, imprecision in reconstructing spatial high-frequency components in a small region might be masked by the picture's local "busyness." On the other hand, highly precise representation and reconstruction of the average value or dc term of the DCT block would be important in a smooth area of sky. The dc $F(0,0)$ term of the transformed coefficients represents the average of the original 64 coefficients.

As stated previously, the DCT of each 8 × 8 block of pixel values produces an 8 × 8 array of DCT coefficients. The relative precision accorded to each of the 64 DCT coefficients can be selected according to its relative importance in human visual perception. The relative coefficient precision information is represented by a *quantizer matrix*, which is an 8×8 array of values. Each value in the quantizer matrix represents the coarseness of quantization of the related DCT coefficient.

Two types of quantizer matrices are supported:

- A matrix used for macroblocks that are intraframe-coded

- A matrix used for non-intraframe-coded macroblocks

The video-coding system defines default values for both the intraframe-quantizer and the non-intraframe-quantizer matrices. Either or both of the quantizer matrices can be overridden at the picture level by transmission of appropriate arrays of 64 values. Any quantizer matrix overrides stay in effect until the following sequence start code.

The transform coefficients, which represent the bulk of the actual coded video information, are quantized to various degrees of coarseness. As indicated previously, some portions of the picture will be more affected in appearance than others by the loss of precision through coefficient quantization. This phenomenon is exploited by the availability of the quantizer scale factor, which allows the overall level of quantization to vary for each macroblock. Consequently, entire macroblocks that are deemed to be visually less important can be quantized more coarsely, resulting in fewer bits being needed to represent the picture.

For each coefficient other than the dc coefficient of intraframe-coded blocks, the quantizer scale factor is multiplied by the corresponding value in the appropriate quantizer matrix to form the quantizer step size. Quantization of the dc coefficients of intraframe-coded blocks is unaffected by the quantizer scale factor and is governed only by the (0, 0) element of the intraframe-quantizer matrix, which always is set to be 8 (ISO/IEC 13818-2).

Entropy Coder

An important effect of the quantization of transform coefficients is that many coefficients will be rounded to zero after quantization [10]. In fact, a primary method of controlling the encoded data rate is the control of quantization coarseness, because a coarser quantization leads to an increase in the number of zero-value quantized coefficients.

Inverse Quantizer

At the decoder, the coded coefficients are decoded, and an 8×8 block of quantized coefficients is reconstructed [10]. Each of these 64 coefficients is *inverse-quantized* according to the prevailing quantizer matrix, quantizer scale, and frame type. The result of inverse quantization is a block of 64 DCT coefficients.

Inverse Spatial Transform Block—IDCT

The decoded and inverse-quantized coefficients are organized as 8×8 blocks of DCT coefficients, and the inverse discrete cosine transform is applied to each block [10]. This results in a new array of pixel values, or pixel difference values that correspond to the output of the subtraction at the beginning of the prediction loop. If the prediction loop was in the interframe mode, the values will be pixel differences. If the loop was in the intraframe mode, the inverse transform will produce pixel values directly.

Motion Compensator

If a portion of the image has not moved, then it is easy to see that a subtraction of the old portion from the new portion of the image will produce zero or nearly zero pixel differences, which is the goal of the prediction [10]. If there has been movement in the portion of the image under consideration, however, the direct pixel-by-pixel differences generally will not be zero, and might be statistically very large. The motion in most natural scenes is organized, however, and can be approximately represented locally as a translation in most cases. For this reason, the video-coding system allows for *motion-compensated* prediction, whereby macroblock-sized regions in the reference frame may be translated vertically and horizontally with respect to the macroblock being predicted, to compensate for local motion.

The pixel-by-pixel differences between the current macroblock and the motion-compensated prediction are transformed by the DCT and quantized using the composition of the non-intraframe-quantizer matrix and the quantizer scale factor. The quantized coefficients then are coded.

2.5.8 Dual Prime Prediction Mode

The dual prime prediction mode is an alternative "special" prediction mode that is built on field-based motion prediction but requires fewer transmitted motion vectors than conventional field-based prediction [10]. This mode of prediction is available only for interlaced material and only when the encoder configuration does not use *B*-frames. This mode of prediction can be particularly useful for improving encoder efficiency for low-delay applications.

The basis of dual prime prediction is that field-based predictions of both fields in a macroblock are obtained by averaging two separate predictions, which are predicted from the two nearest decoded fields in time. Each of the macroblock fields is predicted separately, although the four vectors (one pair per field) used for prediction all are derived from a single transmitted field-based motion vector. In addition to the single field-based motion vector, a small *differential vector* (limited to vertical and horizontal component values of +1, 0, and –1) also is transmitted for each macroblock. Together, these vectors are used to calculate the pairs of motion vectors for each macroblock. The first prediction in the pair is simply the transmitted field-based motion vector. The second prediction vector is obtained by combining the differential vector with a scaled version of the first vector. After both predictions are obtained, a single prediction for each macroblock field is calculated by averaging each pixel in the two original predictions. The final averaged prediction then is subtracted from the macroblock field being encoded.

2.5.9 Adaptive Field/Frame Prediction Mode

Interlaced pictures may be coded in one of two ways: either as two separate fields or as a single frame [10]. When the picture is coded as separate fields, all of the codes for the first field are transmitted as a unit before the codes for the second field. When the picture is coded as a frame, information for both fields is coded for each macroblock.

When frame-based coding is used with interlaced pictures, each macroblock may be selectively coded using either field prediction or frame prediction. When frame prediction is used, a motion vector is applied to a picture region that is made up of both parity fields interleaved together. When field prediction is used, a motion vector is applied to a region made up of scan lines from a single field. Field prediction allows the selection of either parity field to be used as a reference for the field being predicted.

2.5.10 Image Refresh

As discussed previously, a given picture may be sent by describing the differences between it and one or two previously transmitted pictures [10]. For this scheme to work, there must be some way for decoders to become initial-ized with a valid picture upon tuning into a new channel, or to become reini-tialized with a valid picture after experiencing transmission errors. Additionally, it is necessary to limit the number of consecutive predictions that can be performed in a decoder to control the buildup of errors resulting from *IDCT mismatch*.

IDCT mismatch occurs because the video-coding system, by design, does not completely specify the results of the IDCT operation. MPEG did not fully specify the results of the IDCT to allow for evolutionary improvements in implementations of this computationally intensive operation. As a result, it is possible for the reconstructed pictures in a decoder to "drift" from those in the encoder if many successive predictions are used, even in the absence of transmission errors. To control the amount of drift, each macroblock is required to be coded without prediction (intraframe-coded) at least once in any 132 consecutive frames.

The process whereby a decoder becomes initialized or reinitialized with valid picture data—without reference to previously transmitted picture infor-mation—is termed *image refresh*. Image refresh is accomplished by the use of intraframe-coded macroblocks. The two general classes of image refresh, which can be used either independently or jointly, are:

- Periodic transmission of *I*-frames

- Progressive refresh

Periodic Transmission of I-Frames

One simple approach to image refresh is to periodically code an entire frame using only intraframe coding [10]. In this case, the intra-coded frame is typi-cally an *I*-frame. Although prediction is used within the frame, no reference is made to previously transmitted frames. The period between successive intracoded frames may be constant, or it may vary. When a receiver tunes into a new channel where *I*-frame coding is used for image refresh, it may perform the following steps:

- Ignore all data until receipt of the first sequence header

- Decode the sequence header, and configure circuits based on sequence parameters

- Ignore all data until the next received *I*-frame

- Commence picture decoding and presentation

When a receiver processes data that contains uncorrectable errors in an *I*-
or *P*-frame, there typically will be a propagation of picture errors as a result
of predictive coding. Pictures received after the error may be decoded incorrectly until an error-free *I*-frame is received.

Progressive Refresh

An alternative method for accomplishing image refresh is to encode only a
portion of each picture using the intraframe mode [10]. In this case, the
intraframe-coded regions of each picture should be chosen in such a way
that, over the course of a reasonable number of frames, all macroblocks are
coded intraframe at least once. In addition, constraints might be placed on
motion-vector values to avoid possible contamination of refreshed regions
through predictions using unrefreshed regions in an uninitialized decoder.

2.5.11 Discrete Cosine Transform

Predictive coding in the MPEG-2 compression algorithm exploits the temporal correlation in the sequence of image frames [10]. Motion compensation is a refinement of that temporal prediction, which allows the coder to
account for apparent motions in the image that can be estimated. Aside from
temporal prediction, another source of correlation that represents redundancy in the image data is the spatial correlation within an image frame or
field. This spatial correlation of images, including parts of images that contain apparent motion, can be accounted for by a spatial transform of the prediction differences. In the intraframe-coding case, where there is by
definition no attempt at prediction, the spatial transform applies to the actual
picture data. The effect of the spatial transform is to concentrate a large fraction of the signal energy in a few transform coefficients.

To exploit spatial correlation in intraframe and predicted portions of the
image, the image-prediction residual pixels are represented by their DCT
coefficients. For typical images, a large fraction of the energy is concen-
trated in a few of these coefficients. This makes it possible to code only a
few coefficients without seriously affecting the picture quality. The DCT is
used because it has good energy-compaction properties and results in real
coefficients. Furthermore, numerous fast computational algorithms exist for
implementation of DCT.

Blocks of 8×8 Pixels

Theoretically, a large DCT will outperform a small DCT in terms of coefficient decorrelation and block energy compaction [10]. Better overall performance can be achieved, however, by subdividing the frame into many smaller regions, each of which is individually processed.

If the DCT of the entire frame is computed, the whole frame is treated equally. For a typical image, some regions contain a large amount of detail, and other regions contain very little. Exploiting the changing characteristics of different images and different portions of the same image can result in significant improvements in performance. To take advantage of the varying characteristics of the frame over its spatial extent, the frame is partitioned into blocks of 8×8 pixels. The blocks then are independently transformed and adaptively processed based on their local characteristics. Partitioning the frame into small blocks before taking the transform not only allows spatially adaptive processing, but also reduces the computational and memory requirements. The partitioning of the signal into small blocks before computing the DCT is referred to as the *block DCT*.

An additional advantage of using the DCT domain representation is that the DCT coefficients contain information about the spatial frequency content of the block. By utilizing the spatial frequency characteristics of the human visual system, the precision with which the DCT coefficients are transmitted can be in accordance with their perceptual importance. This is achieved through the quantization of these coefficients, as explained in the following sections.

Adaptive Field/Frame DCT

As noted previously, the DCT makes it possible to take advantage of the typically high degree of spatial correlation in natural scenes [10]. When interlaced pictures are coded on a frame basis, however, it is possible that significant amounts of motion result in relatively low spatial correlation in some regions. This situation is accommodated by allowing the DCTs to be computed either on a field basis or on a frame basis. The decision to use field- or frame-based DCT is made individually for each macroblock.

Adaptive Quantization

The goal of video compression is to maximize the video quality at a given bit rate, and this requires a careful distribution of the limited number of available bits [10]. By exploiting the perceptual irrelevancy and statistical redundancy within the DCT domain representation, an appropriate bit allocation can yield significant improvements in performance. Quantization is performed to reduce the precision of the DCT coefficient values, and through quantization and code word assignment, the actual bit-rate compression is achieved. The quantization process is the source of virtually all the loss of information in the compression algorithm. This is an important point, as it simplifies the design process and facilitates fine-tuning of the system.

The degree of subjective picture degradation caused by coefficient quantization tends to depend on the nature of the scenery being coded. Within a given picture, distortions of some regions may be less apparent than in others. The video-coding system allows for the level of quantization to be adjusted for each macroblock in order to save bits, where possible, through coarse quantization.

Perceptual Weighting

The human visual system is not uniformly sensitive to coefficient quantization error [10]. Perceptual weighting of each source of coefficient quantization error is used to increase quantization coarseness, thereby lowering the bit rate. The amount of visible distortion resulting from quantization error for a given coefficient depends on the coefficient number, or frequency, the local brightness in the original image, and the duration of the temporal characteristic of the error. The dc coefficient error results in mean value distortion for the corresponding block of pixels, which can expose block boundaries. This is more visible than higher-frequency coefficient error, which appears as noise or texture.

Displays and the HVS exhibit nonuniform sensitivity to detail as a function of local average brightness. Loss of detail in dark areas of the picture is not as visible as it is in brighter areas. Another opportunity for bit savings is presented in textured areas of the picture, where high-frequency coefficient error is much less visible than in relatively flat areas. Brightness and texture weighting require analysis of the original image because these areas may be well predicted. Additionally, distortion can be easily masked by limiting its duration to one or two frames. This effect is most profitably used after scene changes, where the first frame or two can be greatly distorted without perceptible artifacts at normal speed.

When transform coefficients are being quantized, the differing levels of perceptual importance of the various coefficients can be exploited by "allocating the bits" to shape the quantization noise into the perceptually less important areas. This can be accomplished by varying the relative step sizes of the quantizers for the different coefficients. The perceptually important coefficients may be quantized with a finer step size than the others. For example, low spatial frequency coefficients may be quantized finely, and the less important high-frequency coefficients may be quantized more coarsely. A simple method to achieve different step sizes is to normalize or weight each coefficient based on its visual importance. All of the normalized coefficients may then be quantized in the same manner, such as rounding to the nearest integer (uniform quantization). Normalization or weighting effectively scales the quantizer from one coefficient to another. The MPEG-2 video-compression system utilizes perceptual weighting, where the different DCT coefficients are weighted according to a perceptual criterion prior to uniform quantization. The perceptual weighting is determined by quantizer matrices. The compression system allows for modifying the quantizer matrices before each picture.

2.5.12 Entropy Coding of Video Data

Quantization creates an efficient, discrete representation for the data to be transmitted [10]. Code word assignment takes the quantized values and produces a digital bit stream for transmission. Hypothetically, the quantized values could be simply represented using uniform- or fixed-length code words. Under this approach, every quantized value would be represented with the same number of bits. As outlined previously in general terms, greater efficiency—in terms of bit rate—can be achieved with entropy coding.

Entropy coding attempts to exploit the statistical properties of the signal to be encoded. A signal, whether it is a pixel value or a transform coefficient, has a certain amount of information, or entropy, based on the probability of the different possible values or events occurring. For example, an event that occurs infrequently conveys much more new information than one that occurs often. The fact that some events occur more frequently than others can be used to reduce the average bit rate.

Huffman Coding

Huffman coding, which is utilized in the ATSC DTV video-compression system, is one of the most common entropy-coding schemes [10]. In Huffman coding, a code book is generated that can approach the minimum average description length (in bits) of events, given the probability distribution of all the events. Events that are more likely to occur are assigned shorter-length code words, and those less likely to occur are assigned longer-length code words.

Run Length Coding

In video compression, most of the transform coefficients frequently are quantized to zero [10]. There may be a few non-zero low-frequency coefficients and a sparse scattering of non-zero high-frequency coefficients, but most of the coefficients typically have been quantized to zero. To exploit this phenomenon, the 2-dimensional array of transform coefficients is reformatted and prioritized into a 1-dimensional sequence through either a zigzag- or alternate-scanning process. This results in most of the important non-zero coefficients (in terms of energy and visual perception) being grouped together early in the sequence. They will be followed by long runs of coefficients that are quantized to zero. These zero-value coefficients can be efficiently represented through *run length encoding*.

In run length encoding, the number (run) of consecutive zero coefficients before a non-zero coefficient is encoded, followed by the non-zero coefficient value. The run length and the coefficient value can be entropy-coded, either separately or jointly. The scanning separates most of the zero and the non-zero coefficients into groups, thereby enhancing the efficiency of the run length encoding process. Also, a special *end-of-block* (EOB) marker is used to signify when all of the remaining coefficients in the sequence are equal to zero. This approach can be extremely efficient, yielding a significant degree of compression.

In the alternate-/zigzag-scan technique, the array of 64 DCT coefficients is arranged in a 1-dimensional vector before run length/amplitude code word assignment. Two different 1-dimensional arrangements, or *scan types*, are allowed, generally referred to as *zigzag scan* (shown in Figure 2.11*a*) and *alternate scan* (shown in Figure 2.11*b*). The scan type is specified before coding each picture and is permitted to vary from picture to picture.

Channel Buffer

Whenever entropy coding is employed, the bit rate produced by the encoder is variable and is a function of the video statistics [10]. Because the bit rate permitted by the transmission system is less than the peak bit rate that may be produced by the variable-length coder, a *channel buffer* is necessary at the decoder. This buffering system must be carefully designed. The buffer con-

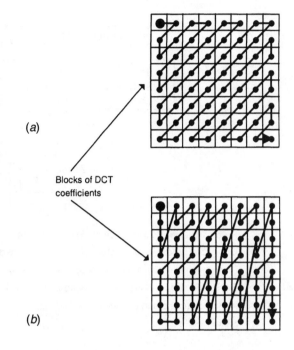

(a)

Blocks of DCT
coefficients

(b)

Figure 2.11 Scanning of coefficient blocks: (*a*) alternate scanning of coefficients, (*b*) zigzag scanning of coefficients. (*From* [10]. *Used with permission.*)

troller must allow efficient allocation of bits to encode the video and also ensure that no overflow or underflow occurs.

Buffer control typically involves a feedback mechanism to the compression algorithm whereby the amplitude resolution (quantization) and/or spatial, temporal, and color resolution may be varied in accordance with the instantaneous bit-rate requirements. If the bit rate decreases significantly, a finer quantization can be performed to increase it.

The ATSC DTV standard specifies a channel buffer size of 8 Mbits. The *model buffer* is defined in the DTV video-coding system as a reference for manufacturers of both encoders and decoders to ensure interoperability. To prevent overflow or underflow of the model buffer, an encoder may maintain measures of buffer occupancy and scene complexity. When the encoder needs to reduce the number of bits produced, it can do so by increasing the general value of the quantizer scale, which will increase picture degradation.

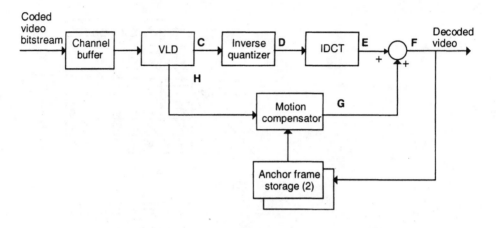

C   Prediction error DCT coefficients in quantized form
D   Quantized prediction error DCT coefficients in standard form
E   Pixel-by-pixel prediction errors, degraded by quantization
F   Reconstructed pixel values, degraded by quantization
G   Motion compensated predicted pixel values
H   Motion vectors

Figure 2.12 ATSC DTV video system decoder functional block diagram. (*From* [10]. *Used with permission.*)

When it is able to produce more bits, it can decrease the quantizer scale, thereby decreasing picture degradation.

Decoder Block Diagram

As shown in Figure 2.12, the ATSC DTV video decoder contains elements that invert, or undo, the processing performed in the encoder [10]. The incoming coded video bit stream is placed in the channel buffer, and bits are removed by a *variable length decoder* (VLD).

The VLD reconstructs 8×8 arrays of quantized DCT coefficients by decoding run length/amplitude codes and appropriately distributing the coefficients according to the scan type used. These coefficients are dequantized and transformed by the IDCT to obtain pixel values or prediction errors.

In the case of interframe prediction, the decoder uses the received motion vectors to perform the same prediction operation that took place in the encoder. The prediction errors are summed with the results of motion-compensated prediction to produce pixel values.

2.5.13 Spatial and S/N Scalability

Because MPEG-2 was designed in anticipation of the need for handling different picture sizes and resolutions, including standard definition television and high-definition television, provisions were made for a hierarchical split of the picture information into a base layer and two enhancement layers [10]. In this way, SDTV decoders would not be burdened with the cost of decoding an HDTV signal.

An encoder for this scenario could work as follows. The HDTV signal would be used as the starting point. It would be spatially filtered and subsampled to create a standard resolution image, which then would be MPEG-encoded. The higher-definition information could be included in an enhancement layer.

Another use of a hierarchical split would be to provide different picture quality without changing the spatial resolution. An encoder quantizer block could realize both coarse and fine filtering levels. Better error correction could be provided for the more coarse data, so that as signal strength weakened, a step-by-step reduction in the picture signal-to-noise ratio would occur in a way similar to that experienced in broadcast analog signals today. Viewers with poor reception, therefore, would experience a more graceful degradation in picture quality instead of a sudden dropout.

2.6   Concatenation

The production of a video segment or program is a serial process: multiple modifications must be made to the original material to yield a finished product. This serial process demands many steps where compression and decompression could take place. Compression and decompression within the same format is not normally considered concatenation. Rather, concatenation involves changing the values of the data, forcing the compression technology to once again compress the signal.

Compressing video is not, generally speaking, a completely lossless process; lossless bit-rate reduction is practical only at the lowest compression ratios. It should be understood, however, that lossless compression is possible—in fact, it is used for critical applications such as medical imaging. Such systems, however, are inefficient in terms of bit usage.

For common video applications, concatenation results in artifacts and coding problems when different compression schemes are cascaded and/or when recompression is required. Multiple generations of coding and decoding are practical, but not particularly desirable. In general, the fewer generations, the better.

Using the same compression algorithm repeatedly (MPEG-2, for example) within a chain—multiple generations, if you will—should not present problems, as long as the pictures are not manipulated (which would force the signal to be recompressed). If, on the other hand, different compression algorithms are cascaded, all bets are off. A detailed mathematical analysis will reveal that such concatenation can result in artifacts ranging from insignificant and unnoticeable to considerable and objectionable, depending on a number of variables, including the following:

- The types of compression systems used

- The compression ratios of the individual systems

- The order or sequence of the compression schemes

- The number of successive coding/decoding steps

- The input signals themselves

The last point merits some additional discussion. Artifacts from concatenation are most likely during viewing of scenes that are difficult to code in the first place, such as those containing rapid movement of objects or noisy signals. Many video engineers are familiar with test tapes containing scenes that are intended specifically to point out the weaknesses of a given compression scheme or a particular implementation of that scheme. To the extent that such scenes represent real-world conditions, these "compression-killer" images represent a real threat to picture quality when subjected to concatenation of systems.

2.6.1 Compression Artifacts

For any video-compression system, the skill of the mathematicians writing the algorithms lies in making the best compromises between preserving the perceived original scene detail and reducing the amount of data actually transmitted. At the limits of these compromises lie artifacts, which vary depending upon the compression scheme employed. Quantifying the degradation is difficult because the errors are subjective: what is obvious to a trained observer may go unnoticed by a typical viewer or by a trained observer under less-than-ideal conditions. Furthermore, such degradation tends to be transient, whereas analog system degradations tend to be constant.

To maintain image quality in a digital system, bottlenecks must be eliminated throughout the signal path. In any system, the signal path is only as good as its weakest element or its worst compression algorithm. It is a logical assumption that the lower the compression ratio, the better the image quality. In fact, however, there is a point of diminishing return, with the increased data simply eating up bandwidth with no apparent quality improvement. These tradeoffs must be made carefully because once picture elements are lost, they cannot be fully recovered

Typical MPEG Artifacts

Although each type of program sequence consists of a unique set of video parameters, certain generalizations can be made with regard to the artifacts that may be expected with MPEG-based compression systems [13]. The artifacts are determined in large part by the algorithm implementations used by specific MPEG encoding vendors. Possible artifacts include the following:

- *Block effects*. These may be seen when the eye tracks a fast-moving, detailed object across the screen. The blocky grid appears to remain fixed while the object moves beneath it. This effect also may be seen during dissolves and fades. It typically is caused by poor motion estimation and/or insufficient allocation of bits in the coder.

- *Mosquito noise*. This artifact may be seen at the edges of text, logos, and other sharply defined objects. The sharp edges cause high-frequency DCT terms, which are coarsely quantized and spread spatially when transformed back into the pixel domain.

- *Dirty window*. This condition appears as streaking or noise that remains stationary while objects move beneath it. In this case, the encoder may not be sending sufficient bits to code the residual (prediction) error in *P*- and *B*-frames.

- *Wavy noise*. This artifact often is seen during slow pans across highly detailed scenes, such as a crowd in a stadium. The coarsely quantized high-frequency terms resulting from such images can cause reconstruction errors to modulate spatially as details shift within the DCT blocks.

It follows, then, that certain types of motion do not fit the MPEG linear translation model particularly well and are, therefore, problematic. These types of motions include:

- Zooms

- Rotations

- Transparent and/or translucent moving objects

- Dissolves containing moving objects

Furthermore, certain types of image elements cannot be predicted well. These image elements include:

- Shadows

- Changes in brightness resulting from fade-ins and fade-outs

- Highly detailed regions

- Noise effects

- Additive noise

Efforts continue to minimize coding artifacts. Success lies in the skill of the system designers in adjusting the many operating parameters of a video encoder. One of the strengths of the MPEG standard is that it allows—and even encourages—diversity and innovation in encoder design.

2.7 Video Encoding Process

The function of any video compression device or system is to provide for efficient storage and/or transmission of information from one location or device to another. The encoding process, naturally, is the beginning point of this chain. Like any chain, video encoding represents not just a single link but many interconnected and interdependent links. The bottom line in video and audio encoding is to ensure that the compressed signal or data stream represents the information required for recording and/or transmission, and *only* that information. If there is additional information of any nature remaining in the data stream, it will take bits to store and/or transmit, which will result in fewer bits being available for the required data. Surplus information is irrelevant because the intended recipient(s) do not require it and can make no use of it.

Surplus information can take many forms. For example, it can be information in the original signal or data stream that exceeds the capabilities of the receiving device to process and display. There is little point in transmitting more resolution than the receiving device can use. Noise is another form of surplus information. Noise is—by nature—random or nearly so, and this makes it essentially incompressible. Many other types of artifacts exist, ranging from filter ringing to film scratches. Some may seem trivial, but in the field of compression they can be very important. Compression relies on order and consistency for best performance, and such artifacts can compromise the final displayed images or at least lower the achievable bit rate reduction. Generally speaking, compression systems are designed for particular tasks, and make use of certain basic assumptions about the nature of the data being compressed.

2.7.1 Encoding Tools

In the migration to digital video technologies, the encoding process has taken on a new and pivotal role. Like any technical advance, however, encoding presents both challenges and rewards. The challenge involves assimilating new tools and new skills. The quality of the final compressed video is dependent upon the compression system used to perform the encoding, the tools provided by the system, and the skill of the person operating the system.

Beyond the automated procedures of encoding lies an interactive process that can considerably enhance the finished video output. These "human-assisted" procedures can make the difference between high-quality images an mediocre ones, and the difference between the efficient use of media and wasted bandwidth.

The goal of intelligent encoding is to minimize the impact of encoding artifacts, rendering them inconspicuous or even invisible. Success is in the eye of the viewer and, thus, the process involves many subjective visual and aesthetic judgments. It is reasonable to conclude, then, that automatic encoding can go only so far. It cannot substitute for the trained eye of the video professional.

In this sense, human-assisted encoding is analogous to the telecine process. In the telecine application, a skilled professional—the *colorist*—uses techniques such as color correction, filtering, and noise reduction to ensure that the video version of a motion picture or other film-based material is as true to the original as technically possible. This work requires a combination of technical expertise and video artistry. Like the telecine, human-assisted encoding is an iterative process. The operator (*compressionist*, if you will) sets the encoding parameters, views the impact of the settings on the scene, and then further modifies the parameters of the encoder until the desired visual result is achieved for the scene or segment.

2.7.2 Signal Conditioning

Correctly used, signal conditioning can provide a remarkable increase in coding efficiency and ultimate picture quality. Encoding equipment available today incorporates many different types of filters targeted at different types of artifacts. The benefits of appropriate conditioning are twofold:

- Because the artifacts are unwanted, there is a clear advantage in avoiding the allocation of bits to transmit them.

- Because the artifacts do not "belong," they generally violate the rules or assumptions of the compression system. For this reason, artifacts do not compress well and use a disproportionately high number of bits to transmit and/or store.

Filtering prior to encoding can be used to selectively screen out image information that might otherwise result in unwanted artifacts. Spatial filtering applies within a particular frame, and can be used to screen out higher frequencies, removing fine texture noise and softening sharp edges. The resulting picture may have a softer appearance, but this is often preferable to a blocking or ringing artifact. Similarly, temporal (recursive) filtering, applied from frame to frame, can be employed to remove temporal noise caused—for example—by grain-based irregularities in film.

Color correction can be used in much the same manner as filtering. Color correction can smooth out uneven areas of color, reducing the amount of data the compression algorithm will have to contend with, thus eliminating artifacts. Likewise, adjustments in contrast and brightness can serve to mask artifacts, achieving some image quality enhancements without noticeably altering the video content.

For decades, the phrase *garbage-in, garbage-out* has been the watchword of the data processing industry. If the input to some process is flawed, the output will invariably be flawed. Such is the case for video compression. Unless proper attention is paid to the entire encoding process, degradations will occur. In general, consider the following encoding checklist:

- Begin with the best. If the source material is film, use the highest-quality print or negative available. If the source is video, use the highest quality, fewest-generation tape.

- Clean up the source material before attempting to compress it. Perform whatever noise reduction, color correction, scratch removal, and other artifact-elimination steps that are possible before attempting to send the signal to the encoder. There are some defects that the encoding process may hide. Noise, scratches, and color errors are not among them. Encoding will only make them worse.

- Decide on an aspect ratio conversion strategy (if necessary). Keep in mind that once information is discarded, it cannot be reclaimed.

- Treat the encoding process like a high-quality telecine transfer. Start with the default compression settings and adjust as needed to achieve the desired result. Document the settings with an *encoding decision list* (EDL) so that the choices made can be reproduced at a later date, if necessary.

The encoding process is much more of an artistic exercise than it is a technical one. In the area of video encoding, there is no substitute for training and experience.

2.7.3 SMPTE RP 202

SMPTE Recommended Practice 202 (proposed at this writing) is an important step in the world of digital video production. Equipment conforming to this practice will minimize artifacts in multiple generations of encoding and de-coding by optimizing macroblock alignment [14]. As MPEG-2 becomes pervasive in emission, contribution, and distribution of video content, multiple compression and decompression (codec) cycles will be required. Concatenation of codecs may be needed for production, post-production, transcoding, or format conversion. Any time video transitions to or from the coefficient domain of MPEG-2 are performed, care must be exercised in alignment of the video, both horizontally and vertically, as it is coded from the raster format or decoded and placed in the raster format.

The first problem is shifting the video horizontally and vertically. Over multiple compression and decompression cycles, this could substantially distort the image. Less obvious, but just as important, is the need for macroblock alignment to reduce artifacts between encoders and decoders from

Table 2.4 Recommended MPEG-2 Coding Ranges for Various Video Formats
(*After* [14].)

Format	Resolution Pels x Lines	Coded Pels	Coded Lines			MPEG-2 Profile and Level
			Field 1	Field 2	Frame	
480I	720 × 480	0–719	23–262	286–525		MP@ML
480P	720 × 480	0–719			46–525	MP@HL
512I	720 × 512	0–719	7–262	270–525		422P@ML
512P	720 × 512	0–719			14–525	422P@HL
576I	720 × 576	0–719	23–310	336–623		MP@ML
608I	720 × 608	0–719	7–310	320–623		422P@ML
720P	1280 × 720	0–1279			26–745	MP@HL
720P	1280 × 720	0–1279			26–745	422P@HL
1080I	1920 × 1088[1]	0–1919	21–560	584–1123		MP@HL
1080I	1920 × 1088[1]	0–1919	21–560	584–1123		422P@HL
1080P	1920 × 1088[1]	0–1919			42–1121	MP@HL
1080P	1920 × 1088[1]	0–1919			42–1121	422P@HL

1 The active image only occupies the first 1080 lines.

various equipment vendors. If concatenated encoders do not share common macroblock boundaries, then additional quantization noise, motion estimation errors, and poor mode decisions may result. Likewise, encoding decisions that may be carried through the production and post-production process with recoding data present, will rely upon macroblock alignment. Decoders must also exercise caution in placement of the active video in the scanning format so that the downstream encoder does not receive an offset image.

With these issues in mind, RP 202 specifies the spatial alignment for MPEG-2 video encoders and decoders. Both standard definition and high-definition video formats for production, distribution, and emission systems are addressed. Table 2.4 gives the recommended coding ranges for MPEG-2 encoders and decoders.

2.8   Digital Audio Data Compression

As with video, high on the list of priorities for the professional audio industry is to refine and extend the range of digital equipment capable of the capture, storage, post production, exchange, distribution, and transmission of high-quality audio, be it mono, stereo, or 5.1 channel AC-3 [15]. This demand being driven by end-users, broadcasters, film makers, and the recording industry alike, who are moving rapidly towards a "tapeless" environment. Over the last two decades, there have been continuing advances in DSP technology, which have supported research engineers in their endeavors to produce the necessary hardware, particularly in the field of digital audio data compression or—as it is often referred to—*bit-rate reduction*. There exist a number of real-time or—in reality—near instantaneous compression coding algorithms. These can significantly lower the circuit bandwidth and storage requirements for the transmission, distribution, and exchange of high-quality audio.

The introduction in 1983 of the compact disc (CD) digital audio format set a quality benchmark that the manufacturers of subsequent professional audio equipment strive to match or improve. The discerning consumer now expects the same quality from radio and television receivers. This leaves the broadcaster with an enormous challenge.

2.8.1 PCM Versus Compression

It can be an expensive and complex technical exercise to fully implement a linear *pulse code modulation* (PCM) infrastructure, except over very short distances and within studio areas [15]. To demonstrate the advantages of distributing compressed digital audio over wireless or wired systems and networks, consider again the CD format as a reference. The CD is a 16 bit linear PCM process, but has one major handicap: the amount of circuit bandwidth the digital signal occupies in a transmission system. A stereo CD transfers information (data) at 1.411 Mbits/s, which would require a circuit with a bandwidth of approximately 700 kHz to avoid distortion of the digital signal. In practice, additional bits are added to the signal for channel coding, synchronization, and error correction; this increases the bandwidth demands yet again. 1.5 MHz is the commonly quoted bandwidth figure for a circuit capable of carrying a CD or similarly coded linear PCM digital stereo signal. This can be compared with the 20 kHz needed for each of two circuits to distribute the same stereo audio in the analog format, a 75-fold increase in bandwidth requirements.

2.8.2 Audio Bit-Rate Reduction

In general, analog audio transmission requires fixed input and output bandwidths [16]. This condition implies that in a real-time compression system, the quality, bandwidth, and distortion/noise level of both the original and the decoded output sound should not be *subjectively* different, thus giving the appearance of a lossless and real-time process.

In a technical sense, all practical real-time bit-rate-reduction systems can be referred to as "lossy." In other words, the digital audio signal at the output is not identical to the input signal data stream. However, some compression algorithms are, for all intents and purposes, lossless; they lose as little as 2 percent of the original signal. Others remove approximately 80 percent of the original signal.

Redundancy and Irrelevancy

A complex audio signal contains a great deal of information, some of which, because the human ear cannot hear it, is deemed irrelevant. [16]. The same signal, depending on its complexity, also contains information that is highly predictable and, therefore, can be made redundant.

Redundancy, measurable and quantifiable, can be removed in the coder and replaced in the decoder; this process often is referred to as *statistical compression*. *Irrelevancy*, on the other hand, referred to as *perceptual coding*, once removed from the signal cannot be replaced and is lost, irretrievably. This is entirely a subjective process, with each proprietary algorithm using a different psychoacoustic model.

Critically perceived signals, such as pure tones, are high in redundancy and low in irrelevancy. They compress quite easily, almost totally a statistical compression process. Conversely, noncritically perceived signals, such as complex audio or noisy signals, are low in redundancy and high in irrelevancy. These compress easily in the perceptual coder, but with the total loss of all the irrelevancy content.

Human Auditory System

The sensitivity of the human ear is biased toward the lower end of the audible frequency spectrum, around 3 kHz [16]. At 50 Hz, the bottom end of the spectrum, and 17 kHz at the top end, the sensitivity of the ear is down by approximately 50 dB relative to its sensitivity at 3 kHz (Figure 2.13). Additionally, very few audio signals—music- or speech-based—carry fundamental frequencies above 4 kHz. Taking advantage of these characteristics of the

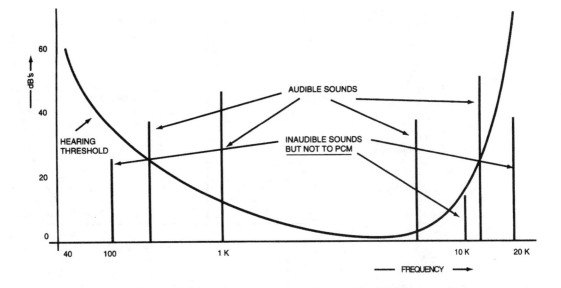

Figure 2.13 Generalized frequency response of the human ear. Note how the PCM process captures signals that the ear cannot distinguish. (*From* [16]. *Used with permission.*)

ear, the structure of audible sounds, and the redundancy content of the PCM signal is the basis used by the designers of the *predictive* range of compression algorithms.

Another well-known feature of the hearing process is that loud sounds mask out quieter sounds at a similar or nearby frequency. This compares with the action of an automatic gain control, turning the gain down when subjected to loud sounds, thus making quieter sounds less likely to be heard. For example, as illustrated in Figure 2.14, if we assume a 1 kHz tone at a level of 70 dBu, levels of greater than 40 dBu at 750 Hz and 2 kHz would be required for those frequencies to be heard. The ear also exercises a degree of temporal masking, being exceptionally tolerant of sharp transient sounds.

It is by mimicking these additional psychoacoustic features of the human ear and identifying the irrelevancy content of the input signal that the *transform* range of low bit-rate algorithms operate, adopting the principle that if the ear is unable to hear the sound then there is no point in transmitting it in the first place.

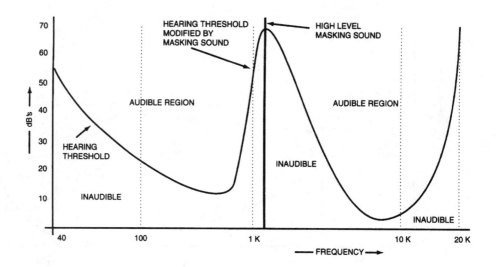

Figure 2.14 Example of the masking effect of a high-level sound. (*From* [16]. *Used with permission.*)

Quantization

Quantization is the process of converting an analog signal to its representative digital format or, as in the case with compression, the requantizing of an already converted signal [16]. This process is the limiting of a finite level measurement of a signal sample to a specific preset integer value. This means that the *actual* level of the sample may be greater or smaller than the preset *reference* level it is being compared with. The difference between these two levels, called the *quantization error*, is compounded in the decoded signal as *quantization noise*.

Quantization noise, therefore, will be injected into the audio signal after each A/D and D/A conversion, the level of that noise being governed by the bit allocation associated with the coding process (i.e., the number of bits allocated to represent the level of each sample taken of the analog signal). For linear PCM, the bit allocation is commonly 16. The level of each audio sample, therefore, will be compared with one of 2^{16} or 65,536 discrete levels or steps.

Compression or bit-rate reduction of the PCM signal leads to the requantizing of an already quantized signal, which will unavoidably inject further quantization noise. It always has been good operating practice to restrict the number of A/D and D/A conversions in an audio chain. Nothing has changed in this regard, and now the number of compression stages also should be kept to a minimum. Additionally, the bit rates of these stages should be set as high as practical; put another way, the compression ratio should be as low as possible.

Sooner or later—after a finite number of A/D, D/A conversions and passes of compression coding, of whatever type—the accumulation of quantization noise and other unpredictable signal degradations eventually will break through the noise/signal threshold, be interpreted as part of the audio signal, be processed as such, and be heard by the listener.

Sampling Frequency and Bit Rate

The bit rate of a digital signal is defined by:

sampling frequency × bit resolution × number of audio channels

The rules regarding the selection of a sampling frequency are based on Nyquist's theorem [16]. This ensures that, in particular, the lower sideband of the sampling frequency does not encroach into the baseband audio. Objectionable and audible aliasing effects would occur if the two bands were to overlap. In practice, the sampling rate is set slightly above twice the highest audible frequency, which makes the filter designs less complex and less expensive.

In the case of a stereo CD with the audio signal having been sampled at 44.1 kHz, this sampling rate produces audio bandwidths of approximately 20 kHz for each channel. The resulting audio bit rate = 44.1 kHz × 16 × 2 = 1.411 Mbits/s, as discussed previously.

2.8.3 Prediction and Transform Algorithms

Most audio-compression systems are based upon one of two basic technologies [16]:

- Predictive or *adaptive differential* PCM (ADPCM) time-domain coding

- Transform or *adaptive* PCM (APCM) frequency-domain coding

It is in their approaches to dealing with the redundancy and irrelevancy of the PCM signal that these techniques differ.

The time domain or *prediction* approach includes G.722, which has been a universal standard since the mid-70s, and was joined in 1989 by a proprietary algorithm, apt-X100. Both these algorithms deal mainly with redundancy.

The frequency domain or *transform* method adopted by a number of algorithms deal in irrelevancy, adopting psychoacoustic masking techniques to identify and remove those unwanted sounds. This range of algorithms include the industry standards ISO/MPEG-1 Layers 1, 2, and 3; apt-Q; MUSICAM; Dolby AC-2 and AC3; and others.

Subband Coding

Without exception, all of the algorithms mentioned in the previous section process the PCM signal by splitting it into a number of frequency subbands, in one case as few as two (G.722) or as many as 1024 (apt-Q) [15]. MPEG-1 Layer 1, with 4:1 compression, has 32 frequency subbands and is the system found in the Digital Compact Cassette (DCC). The MiniDisc ATRAC proprietary algorithm at 5:1 has a more flexible multisubband approach, which is dependent on the complexity of the audio signal.

Subband coding enables the frequency domain redundancies within the audio signals to be exploited. This permits a reduction in the coded bit rate, compared to PCM, for a given signal fidelity. Spectral redundancies are also present as a result of the signal energies in the various frequency bands being unequal at any instant in time. By altering the bit allocation for each subband, either by dynamically adapting it according to the energy of the contained signal or by fixing it for each subband, the quantization noise can be reduced across all bands. This process compares favorably with the noise characteristics of a PCM coder performing at the same overall bit rate.

Subband Gain

On its own, subband coding, incorporating PCM in each band, is capable of providing a performance improvement or *gain* compared with that of full band PCM coding, both being fed with the same complex, constant level input signal [15]. The improvement is defined as *subband gain* and is the ratio of the variations in quantization errors generated in each case while both are operating at the same transmission rate. The gain increases as the number of subbands increase, and with the complexity of the input signal. However, the implementation of the algorithm also becomes more difficult and complex.

Quantization noise generated during the coding process is constrained within each subband and cannot interfere with any other band. The advan-

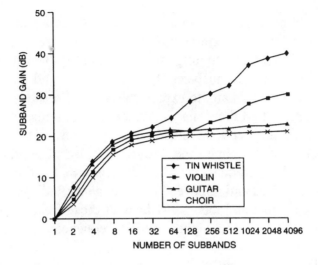

Figure 2.15 Variation of subband gain as a function of the number of subbands. (*From* [16]. *Used with permission.*)

tage of this approach is that the masking by each of the subband dominant signals is much more effective because of the reduction in the noise bandwidth. Figure 2.15 charts subband gain as a function of the number of subbands for four essentially stationary, but differing, complex audio signals.

In practical implementations of compression codecs, several factors tend to limit the number of subbands employed. The primary considerations include:

- The level variation of normal audio signals leading to an averaging of the energy across bands and a subsequent reduction in the coding gain

- The coding or processing delay introduced by additional subbands

- The overall computational complexity of the system

The two key issues in the analysis of a subband framework are:

- Determining the likely improvement associated with additional subbands

- Determining the relationships between subband gain, the number of subbands, and the response of the filter bank used to create those subbands

APCM Coding

The APCM processor acts in a similar fashion to an automatic gain control system, continually making adjustments in response to the dynamics—at all frequencies—of the incoming audio signal [15]. Transform coding takes a time block of signal, analyzes it for frequency and energy, and identifies irrelevant content. Again, to exploit the spectral response of the ear, the frequency spectrum of the signal is divided into a number of subbands, and the most important criteria are coded with a bias toward the more sensitive low frequencies. At the same time, through the use of psychoacoustic masking techniques, those frequencies which it is assumed will be masked by the ear are also identified and removed. The data generated, therefore, describes the frequency content and the energy level at those frequencies, with more bits being allocated to the higher-energy frequencies than those with lower energy.

The larger the time block of signal being analyzed, the better the frequency resolution and the greater the amount of irrelevancy identified. The penalty, however, is an increase in coding delay and a decrease in temporal resolution. A balance has been struck with advances in perceptual coding techniques and psychoacoustic modeling leading to increased efficiency. It is reported in [16] that, with this approach to compression, some 80 percent of the input audio can be removed with acceptable results.

This hybrid arrangement of working with time-domain subbands and simultaneously carrying out a spectral analysis can be achieved by using a *dynamic bit allocation* process for each subband. This subband APCM approach is found in the popular range of software-based MUSICAM, Dolby AC-2, and ISO/MPEG-1 Layers 1 and 2 algorithms. Layer 3—a more complex method of coding and operating at much lower bit rates—is, in essence, a combination of the best functions of MUSICAM and ASPEC, another adaptive transform algorithm. Table 2.5 lists the primary operational parameters for these systems.

Additionally, some of these systems exploit the significant redundancy between stereo channels by using a technique known as *joint stereo coding*. After the common information between left and right channels of a stereo signal has been identified, it is coded only once, thus reducing the bit-rate demands yet again.

Each of the subbands has its own defined *masking threshold*. The output data from each of the filtered subbands is requantized with just enough bit resolution to maintain adequate headroom between the quantization noise and the masking threshold for each band. In more complex coders (e.g., ISO/

Table 2.5 Operational Parameters of Subband APCM Algorithm (*After* [16].)

Coding System	Compression Ratio	Subbands	Bit Rate, kbits/s	A to A Delay, ms[1]	Audio Bandwidth, kHz
Dolby AC-2	6:1	256	256	45	20
ISO Layer 1	4:1	32	384	19	20
ISO Layer 2	Variable	32	192–256	>40	20
IOS Layer 3	12:1	576	128	>80	20
MUSICAM	Variable	32	128–384	>35	20

[1] The total system delay (encoder-to-decoder) of the coding system.

MPEG-1 Layer 3), any spare bit capacity is utilized by those subbands with the greater need for increased masking threshold separation. The maintenance of these signal-to-masking threshold ratios is crucial if further compression is contemplated for any postproduction or transmission process.

2.8.4 Processing and Propagation Delay

As noted previously, the current range of popular compression algorithms operate—for all intents and purposes—in real time [15]. However, this process does of necessity introduce some measurable delay into the audio chain. All algorithms take a finite time to analyze the incoming signal, which can range from a few milliseconds to tens and even hundreds of milliseconds. The amount of processing delay will be crucial if the equipment is to be used in any interactive or two-way application. As a rule of thumb, any more than 20 ms of delay in a two-way audio exchange is problematic. Propagation delay in satellite and long terrestrial circuits is a fact of life. A two-way hook up over a 1000 km, full duplex, telecom digital link has a propagation delay of 3 ms in each direction. This is comparable to having a conversation with someone standing 1 m away. It is obvious that even over a very short distance, the use of a codec with a long processing delay characteristic will have a dramatic effect on operation.

2.8.5 Bit Rate and Compression Ratio

The ITU has recommend the following bit rates when incorporating data compression in an audio chain [15]:

- 128 kbits/s per mono channel (256 kbits/s for stereo) as the minimum bit rate for any stage if further compression is anticipated or required.

- 192 kbits/s per mono channel (384 kbits/s for stereo) as the minimum bit rate for the first stage of compression in a complex audio chain.

These markers place a 4:1 compression ratio at the "safe" end in the scale. However, more aggressive compression ratios, currently up to a nominal 20:1, are available. Keep in mind, though, that low bit rate, high-level compression can lead to problems if any further stages of compression are required or anticipated.

With successive stages of compression, either or both the noise floor and the audio bandwidth will be set by the stage operating at the lowest bit rate. It is, therefore, worth emphasizing that after these platforms have been set by a low bit rate stage, they cannot be subsequently improved by using a following stage operating at a higher bit rate.

Bit Rate Mismatch

A stage of compression may well be followed in the audio chain by another digital stage, either of compression or linear, but—more importantly—operating at a different sampling frequency [15]. If a D/A conversion is to be avoided, a sample rate converter must be used. This can be a stand alone unit or it may already be installed as a module in existing equipment. Where a following stage of compression is operating at the same sampling frequency but a different compression ratio, the bit resolution will change by default.

If the stages have the same sampling frequencies, a direct PCM or AES/EBU digital link can be made, thus avoiding the conversion to the analog domain.

2.8.6 Editing Compressed Data

The linear PCM waveform associated with standard audio workstations is only useful if decoded [15]. The resolution of the compressed data may or may not be adequate to allow direct editing of the audio signal. The minimum audio sample that can be removed or edited from a transform-coded signal will be determined by the size of the time block of the PCM signal being analyzed. The larger the time block, the more difficult the editing of the compressed data becomes.

2.8.7 Common Compression Techniques

Subband APCM coding has found numerous applications in the professional audio industry, including [16]:

- The digital compact cassette (DCC)—uses the simplest implementation of subband APCM with the PASC/ISO/MPEG-1 Layer 1 algorithm incorporating 32 subbands offering 4:1 compression and producing a bit rate of 384 kbits/s.

- The MiniDisc with the proprietary ATRAC algorithm—produces 5:1 compression and 292 kbits/s bit rate. This algorithm uses a *modified discrete cosine transform* (MDCT) technique ensuring greater signal analysis by processing time blocks of the signal in nonuniform frequency divisions, with fewer divisions being allocated to the least sensitive higher frequencies.

- ISO/MPEG-1 Layer 2 (MUSICAM by another name)—a software-based algorithm that can be implemented to produce a range of bit rates and compression ratios commencing at 4:1.

- The ATSC DTV system—uses the subband APCM algorithm in Dolby AC-3 for the audio surround system associated with the ATSC DTV standard. AC-3 delivers five audio channels plus a bass-only effects channel in less bandwidth than that required for one stereo CD channel. This configuration is referred to as 5.1 channels.

For the purposes of illustration, two commonly used audio compression systems will be examined in some detail:

- apt-X100

- ISO/MPEG-1 Layer 2

apt-X100

apt-X100 is a four subband prediction (ADPCM) algorithm [15]. Differential coding reduces the bit rate by coding and transmitting or storing only the difference between a predicted level for a PCM audio sample and the absolute level of that sample, thus exploiting the redundancy contained in the PCM signal.

Audio exhibits relatively slowly varying energy fluctuations with respect to time. Adaptive differential coding, which is dependent on the energy of the input signal, dynamically alters the step size for each quantizing interval to reflect these fluctuations. In apt-X100, this equates to the *backwards adaptation process* and involves the analysis of 122 previous samples. Being a continuous process, this provides an almost constant and optimal signal-to-quantization noise ratio across the operating range of the quantizer.

Time domain subband algorithms implicitly model the hearing process and indirectly exploit a degree of irrelevancy by accepting that the human ear is more sensitive at lower frequencies. This is achieved in the four subband derivative by allocating more bits to the lower frequency bands. This is the only application of psychoacoustics exercised in apt-X100. All the information contained in the PCM signal is processed, audible or not (i.e., no attempt is made to remove irrelevant information). It is the unique fixed allocation of bits to each of the four subbands, coupled with the filtering characteristics of each individual listeners' hearing system, that achieves the satisfactory audible end result.

The user-defined output bit rates range from 56 to 384 kbits/s, achieved by using various sampling frequencies from 16 kHz to 48 kHz, which produce audio bandwidths from 7.5 kHz mono to 22 kHz stereo.

Auxiliary data up to 9.6 kbits/s can also be imbedded into the data stream without incurring a bit overhead penalty. When this function is enabled, an audio bit in one of the higher frequency subbands is replaced by an auxiliary data bit, again with no audible effect.

An important feature of this algorithm is its inherent robustness to random bit errors. No audible distortion is apparent for normal program material at a *bit error rate* (BER) of 1:10,000, while speech is still intelligible down to a BER of 1:10.

Distortions introduced by bit errors are constrained within each subband and their impact on the decoder subband predictors and quantizers is proportional to the magnitude of the differential signal being decoded at that instant. Thus, if the signal is small—which will be the case for a low level input signal or for a resonant, highly predictable input signal—any bit error will have minimal effect on either the predictor or quantizer.

The 16 bit linear PCM signal is processed in time blocks of four samples at a time. These are filtered into four equal-width frequency subbands; for 20 kHz, this would be 0–5 kHz, 5–10 kHz, and so on. The four outputs from the *quadrature mirror filter* (QMF) tree are still in the 16-bit linear PCM format, but are now frequency-limited.

As shown in Figure 2.16, the compression process can be mapped by taking, for example, the first and lowest frequency subband. The first step is to create the difference signal. After the system has settled down on initiation,

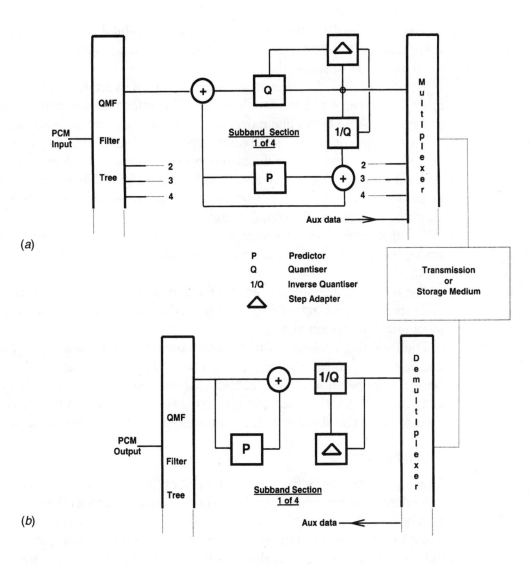

(a)

P   Predictor
Q   Quantiser
1/Q   Inverse Quantiser
△   Step Adapter

(b)

Figure 2.16 apt-X100 audio coding system: (*a*) encoder block diagram, (*b*) decoder block diagram. (*Courtesy of Audio Processing Technology.*)

there will be a reconstructed 16 bit difference signal at the output of the inverse quantizer. This passes into a prediction loop that, having analyzed 122 previous samples, will make a prediction for the level of the next full level sample arriving from the filter tree. This prediction is then compared with the actual level.

The output of the comparator is the resulting 16-bit difference signal. This is requantized to a new 7-bit format, which in turn is inverse quantized back to 16 bits again to enable the prediction loop.

The output from the inverse quantizer is also analyzed for energy content, again for the same 122 previous samples. This information is compared with on-board look up tables and a decision is made to dynamically adjust, up or down as required, the level of each step of the 1024 intervals in the 7-bit quantizer. This ensures that the quantizer will always have adequate range to deal with the varying energy levels of the audio signal. Therefore, the input to the multiplexer will be a 7-bit word but the range of those bits will be varying in relation to the signal energy.

The three other subbands will go through the same process, but the number of bits allocated to the quantizers are much less than for the first subband.

The output of the multiplexer or bit stream formatter is a new 16-bit word that represents four input PCM samples and is, therefore, one quarter of the input rate; a reduction of 4:1.

The decoding process is the complete opposite of the coding procedure. The incoming 16-bit compressed data word is demultiplexed and used to control the operation of four subband decoder sections, each with similar predictor and quantizer step adjusters. A QMF filter tree finally reconstructs a linear PCM signal and separates any auxiliary data that may be present.

ISO/MPEG-1 Layer 2

This algorithm differs from Layer 1 by adopting more accurate quantizing procedures and by additionally removing redundancy and irrelevancy on the generated scale factors [15]. The ISO/MPEG-1 Layer 2 scheme operates on a block of 1152 PCM samples, which at 48 kHz sampling represents a 24 ms time block of the input audio signal. Simplified block diagrams of the encoding/decoding systems are given in Figure 2.17.

The incoming linear PCM signal block is divided into 32 equally spaced subbands using a polyphase analysis filter bank (Figure 2.17a). At 48 kHz sampling, this equates to the bandwidth of each subband being 750 Hz. The bit allocation for the requantizing of these subband samples is then dynamically controlled by information derived from analyzing the audio signal, measured against a preset psychoacoustic model.

The filter bank, which displays manageable delay and minimal complexity, optimally adapts each block of audio to balance between the effects of temporal masking and inaudible pre-echoes.

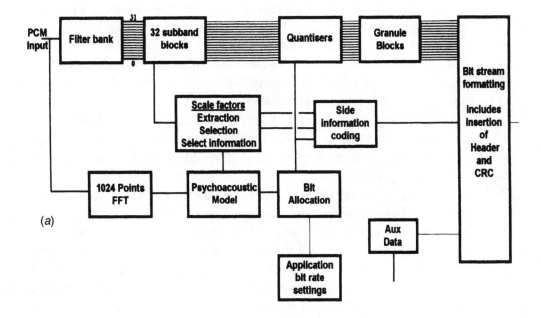

(a)

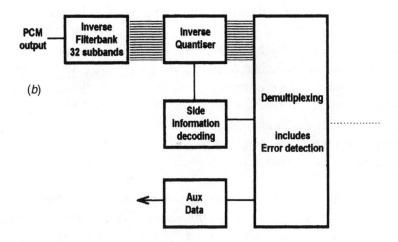

(b)

Figure 2.17 ISO/MPEG-1 Layer 2 system: (*a*) encoder block diagram, (*b*) decoder block diagram. (*After* [15].)

The PCM signal is also fed to a *fast Fourier transform* (FFT) running in parallel with the filter bank. The aural sensitivities of the human auditory system are exploited by using this FFT process to detect the differences between the wanted and unwanted sounds and the quantization noise already present in the signal, and then to adjust the signal-to-mask thresholds, conforming to a preset perceptual model.

This psychoacoustic model is only found in the coder, thus making the decoder less complex and permitting the freedom to exploit future improvements in coder design. The actual number of levels for each quantizer is determined by the bit allocation. This is arrived at by setting the *signal-to-mask ratio* (SMR) parameter, defined as the difference between the minimum masking threshold and the maximum signal level. This minimum masking threshold is calculated using the psychoacoustic model and provides a reference noise level of "just noticeable" noise for each subband.

In the decoder, after demultiplexing and deciphering of the audio and side information data, a dual synthesis filter bank reconstructs the linear PCM signal in blocks of 32 output samples (Figure 2.17*b*).

A scale factor is determined for each 12 subband sample block. The maximum of the absolute values of these 12 samples generates a *scale factor* word consisting of 6 bits, a range of 63 different levels. Because each frame of audio data in Layer 2 corresponds to 36 subband samples, this process will generate 3 scale factors per frame. However, the transmitted data rate for these scale factors can be reduced by exploiting some redundancy in the data. Three successive subband scale factors are analyzed and a pattern is determined. This pattern, which is obviously related to the nature of the audio signal, will decide whether one, two or all three scale factors are required. The decision will be communicated by the insertion of an additional *scale factor select information* data word of 2 bits (SCFSI).

In the case of a fairly stationary tonal-type sound, there will be very little change in the scale factors and only the largest one of the three is transmitted; the corresponding data rate will be $(6 + 2)$ or 8 bits. However, in a complex sound with rapid changes in content, the transmission of two or even three scale factors may be required, producing a maximum bit rate demand of $(6 + 6 + 6 + 2)$ or 20 bits. Compared with Layer 1, this method of coding the scale factors reduces the allocation of data bits required for them by half.

The number of data bits allocated to the overall bit pool is limited or fixed by the data rate parameters. These parameters are set out by a combination of sampling frequency, compression ratio, and—where applicable—the transmission medium. In the case of 20 kHz stereo being transmitted over

ISDN, for example, the maximum data rate is 384 kbits/s, sampling at 48kHz, with a compression ratio of 4:1.

After the number of side information bits required for scale factors, bit allocation codes, CRC, and other functions have been determined, the remaining bits left in the pool are used in the re-coding of the audio subband samples. The allocation of bits for the audio is determined by calculating the SMR, via the FFT, for each of the 12 subband sample blocks. The bit allocation algorithm then selects one of 15 available quantizers with a range such that the overall bit rate limitations are met and the quantization noise is masked as far as possible. If the composition of the audio signal is such that there are not enough bits in the pool to adequately code the subband samples, then the quantizers are adjusted down to a best-fit solution with (hopefully) minimum damage to the decoded audio at the output.

If the signal block being processed lies in the lower one third of the 32 frequency subbands, a 4-bit code word is simultaneously generated to identify the selected quantixer; this word is, again, carried as side information in the main data frame. A 3-bit word would be generated for processing in the mid frequency subbands and a 2-bit word for the higher frequency subbands. When the audio analysis demands it, this allows for *at least* 15, 7, and 3 quantization levels, respectively, in each of the three spectrum groupings. However, each quantizer can, if required, cover from 3 to 65,535 levels and additionally, if no signal is detected then no quantization takes place.

As with the scale factor data, some further redundancy can be exploited, which increases the efficiency of the quantising process. For the lowest quantizer ranges (i.e., 3, 5, and 9 levels), three successive subband sample blocks are grouped into a "granule" and this—in turn—is defined by only one code word. This is particularly effective in the higher frequency subbands where the quantizer ranges are invariably set at the lower end of the scale.

Error detection information can be relayed to the decoder by inserting a 16 bit CRC word in each data frame. This parity check word allows for the detection of up to three single bit errors or a group of errors up to 16 bits in length. A codec incorporating an error concealment regime can either mute the signal in the presence of errors or replace the impaired data with a previous, error free, data frame. The typical data frame structure for ISO/MPEG-1 Layer 2 audio is given in Figure 2.18

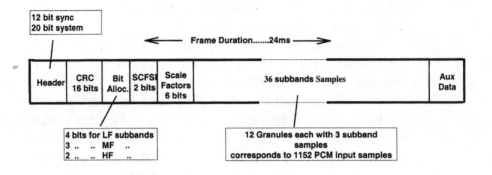

Figure 2.18 ISO/MPEG-1 Layer 2 data frame structure. (*After* [15].)

MPEG-2 AAC

Also of note is MPEG-2 *advanced audio coding* (AAC), a highly advanced perceptual code, used initially for digital radio applications. The AAC code improves on previous techniques to increase coding efficiency. For example, an AAC system operating at 96 kbits/s produces the same sound quality as ISO/MPEG-1 Layer 2 operating at 192 kbits/s—a 2:1 reduction in bit rate. There are three modes (Profiles) in the AAC standard:

- *Main*—used when processing power, and especially memory, are readily available.

- *Low complexity* (LC)—used when processing cycles and memory use are constrained.

- *Scaleable sampling rate* (SSR)—appropriate when a *scalable decoder* is required. A scalable decoder can be designed to support different levels of audio quality from a common bit stream; for example, having both high- and low-cost implementations to support higher and lower audio qualities, respectively.

Different Profiles trade off encoding complexity for audio quality at a given bit rate. For example, at 128 kbits/s, the Main Profile AAC code has a more complex encoder structure than the LC AAC code at the same bit rate, but provides better audio quality as a result.

A block diagram of the AAC system general structure is given in Figure 2.19. The blocks in the drawing are referred to as "tools" that the coding alogrithm uses to compress the digital audio signal. While many of these

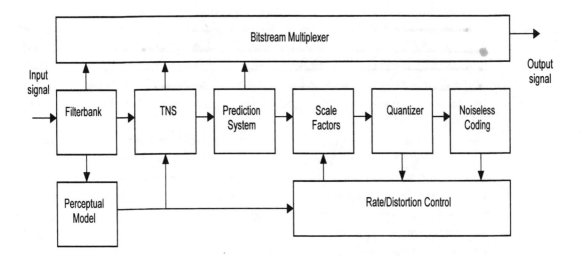

Figure 2.19 Functional block diagram of the MPEG-2 AAC coding system.

tools exist in most audio perceptual codes, two are unique to AAC—the *temporal noise shaper* (TNS) and the *filterbank* tool. The TNS uses a backward adaptive prediction process to remove redundancy between the frequency channels that are created by the filterbank tool.

MPEG-2 AAC provides the capability of up to 48 main audio channels, 16 low frequency effects channels, 16 overdub/multilingual channels, and 10 data streams. By comparison, ISO/MPEG-1 Layer 1 provides two channels and Layer 2 provides 5.1 channels (maximum). AAC is not backward compatible with the Layer 1 and Layer 2 codes.

MPEG-4

MPEG-4, as with the MPEG-1 and MPEG-2 efforts, is not concerned solely with the development of audio coding standards, but also encompasses video coding and data transmission elements (as discussed previously in this chapter). In addition to building upon the audio coding standards developed for MPEG-2, MPEG-4 includes a revolutionary new element—synthesized sound. Tools are provided within MPEG-4 for coding of both natural sounds (speech and music) and for synthesizing sounds based on structured descriptions. The representations used for synthesizing sounds can be formed by text or by instrument descriptions, and by coding other parameters to provide for effects, such as reverberation and spatialization.

Natural audio coding is supported within MPEG-4 at bit rates ranging from 2–64 kbits/s, and includes the MPEG-2 AAC standard (among others) to provide for general compression of audio in the upper bit rate range (8–64 kbits/s), the range of most interest to broadcasters. Other types of coders, primarily voice coders (or *vocoders*) are used to support coding down to the 2 kbits/s rate.

For synthesized sounds, decoders are available that operate based on so-called *structured inputs*, that is, input signals based on descriptions of sounds and not the sounds themselves. Text files are one example of a structured input. In MPEG-4, text can be converted to speech in a *text-to-speech* (TTS) decoder. Synthetic music is another example, and may be delivered at extremely low bit rates while still describing an exact sound signal. The standard's *structured audio decoder* uses a language to define an orchestra made up of instruments, which can be downloaded in the bit stream, not fixed in the decoder.

TTS support is provided in MPEG-4 for unembellished text, or text with prosodic (pitch contour, phoneme duration, etc.) parameters, as an input to generate intelligible synthetic speech. It includes the following functionalities:

- Speech synthesis using the prosody of the original speech

- Facial animation control with phoneme information (important for multimedia applications)

- *Trick mode* functionality: pause, resume, jump forward, jump backward

- International language support for text

- International symbol support for phonemes

- Support for specifying the age, gender, language, and dialect of the speaker

MPEG-4 does not standardize a method of synthesis, but rather a method of describing synthesis.

2.8.8 Dolby E Coding System

Dolby E coding was developed to expand the capacity of existing two channel AES/EBU digital audio infrastructures to make them capable of carrying up to eight channels of audio plus the metadata required by the Dolby Digital coders used in the ATSC transmission system [17]. This allows existing digital videotape recorders, routing switchers, and other video plant equipment, as well as satellite and telco facilities, to be used in program contribution and distribution systems that handle multichannel audio. The coding system was designed to provide broadcast quality output even when decoded and re-encoded many times, and to provide clean transitions when switching between programs.

Dolby E encodes up to eight audio channels plus the necessary metadata and inserts this information into the payload space of a single AES digital audio pair. Because the AES protocol is used as the transport mechanism for the Dolby E encoded signal, digital VTRs, routing switchers, DAs, and all other existing digital audio equipment in a typical video facility can handle multichannel programming. It is possible to do insert or assemble edits on tape or to make audio-follow-video cuts between programs because the Dolby E data is synchronized with the accompanying video. The metadata is multiplexed into the compressed audio, so it is switched with and stays in sync with the audio.

The main challenge in designing a bit-rate reduction system for multiple generations is to prevent coding artifacts from appearing in the recovered audio after several generations. The coding artifacts are caused by a buildup of noise during successive encoding and decoding cycles, so the key to good multigeneration performance is to manage the noise optimally.

This noise is caused by the rate reduction process itself. Digitizing (quantizing) a signal leads to an error that appears in the recovered signal as a broadband noise. The smaller the quantizer steps (i.e., the more resolution or bits used to quantize the signal), the lower the noise will be. This quantizing noise is related to the signal, but becomes "whiter" as the quantizer resolution rises. With resolutions less than about 5 or 6 bits and no dither, the quantizing noise is clearly related to the program material.

Bit rate reduction systems try to squeeze the data rates down to the equivalent of a few bits (or less) per sample and, thus, tend to create quantizing noise in quite prodigious quantities. The key to recovering signals that are subjectively indistinguishable from the original signals, or in which the quantizing noise is inaudible, is in allocating the available bits to the program signal components in a way that takes advantage of the ear's natural ability to mask low level signals with higher level ones.

The rate reduction encoder sends information about the frequency spectrum of the program signal to the decoder. A set of reconstruction filters in

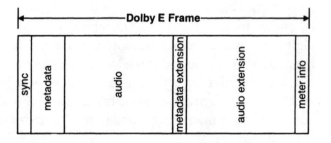
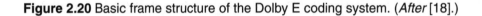

Figure 2.20 Basic frame structure of the Dolby E coding system. (*After* [18].)

the decoder confines the quantizing noise produced by the bit allocation process in the encoder to the bandwidth of those filters. This allows the system designer to keep the noise (ideally) below the masking thresholds produced by the program signal. The whole process of allocating different numbers of bits to different program signal components (or of quantizing them at different resolutions) creates a noise floor that is related to the program signal and to the rate reduction algorithm used. The key to doing this is to have an accurate model of the masking characteristics of the ear, and in allocating the available bits to each signal component so that the masking threshold is not exceeded.

When a program is decoded and then re-encoded, the re-encoding process (and any subsequent ones) adds its noise to the noise already present. Eventually, the noise present in some part of the spectrum will build up to the point where it becomes audible, or exceeds the allowable *coding margin*. A codec designed for minimum data rate has to use lower coding margins (or more aggressive bit allocation strategies) than one intended to produce high quality signals after many generations

The design strategy for a multigeneration rate reduction system, such as one used for Dolby E, is therefore quite different than that of a minimum data rate codec intended for program transmission applications.

Dolby E signals are carried in the AES3 interface using a packetized structure [18]. The packets are based on the coded Dolby E frame, which is illustrated in Figure 2.20. Each Dolby E frame consists of a *synchronization field, metadata field, coded audio field,* and a *meter field*. The metadata field contains a complete set of parameters so that each Dolby E frame can be decoded independently. The Dolby E frames are embedded into the AES3

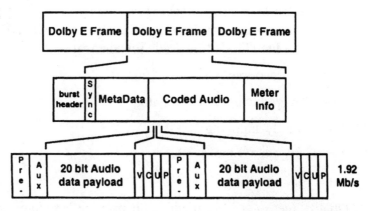

Figure 2.21 Overall coding scheme of Dolby E. (*After* [18].)

interface by mapping the Dolby E data into the audio sample word bits of the AES3 frames utilizing both channels within the signal. (See Figure 2.21.) The data can be packed to utilize 16, 20, or 24 bits in each AES3 sub-frame. The advantage of utilizing more bits per sub-frame is that a higher data rate is available for carrying the coded information. With a 48 kHz AES3 signal, the 16 bit mode allows a data rate of up to 1.536 Mbits/s for the Dolby E signal, while the 20 bit mode allows 1.92 Mbits/s. Higher data rate allows more generations and/or more channels of audio to be supported. However, some AES3 data paths may be restricted in data rate (e.g., some storage devices will only record 16 or 20 bits). Dolby E therefore allows the user to choose the optimal data rate for a given application.

2.8.9 Objective Quality Measurements

Perceptual audio coding has revolutionized the processing and distribution of digital audio signals. One aspect of this technology, not often emphasized, is the difficulty of determining, *objectively*, the quality of perceptually coded signals. Audio professionals could greatly benefit from an objective approach to signal characterization because it would offer a simple but accurate approach for verification of good audio quality within a given facility.

Most of the discussions regarding this topic involve reference to the results of subjective evaluations of audio quality, where for example, groups of listeners compare reference audio material to coded audio material and then judge the *level of impairment* caused by the coding process. A procedure for this process has been standardized in ITU-R Rec. BS.1116, and makes use of the ITU-R five grade impairment scale:

- 5.0—Imperceptible
- 4.0—Perceptible but not annoying
- 3.0—Slightly annoying
- 2.0—Annoying
- 1.0—Very annoying

Quality measurements made with properly executed subjective evaluations are widely accepted and have been used for a variety of purposes, from determining which of a group of perceptual coders performs best, to assessing the overall performance of an audio broadcasting system.

The problem with subjective evaluations is that, while accurate, they are time consuming and expensive to undertake. Traditional objective benchmarks of audio performance, such as signal-to-noise ratio or total harmonic distortion, are not reliable measures of perceived audio quality, especially when perceptually coded signals are being considered.

To remedy this situation, ITU-R established Task Group 10-4 to develop a method of objectively assessing perceived audio quality. Conceptually, the result of this effort would be a device having two inputs—a reference and the audio signal to be evaluated—and would generate an audio quality estimate based on these sources.

Six organizations proposed models for accomplishing this objective, and over the course of several years these models were evaluated for effectiveness, in part by using source material from previously documented subjective evaluations. Ultimately, the task group decided that none of the models by themselves fully met the stated requirements. The group decided, instead, to use the best parts of the different models to create another model that would meet the sought-after requirements.

This approach resulted in an objective measurement method known as *Perceptual Evaluation of Audio Quality* (PEAQ). The method contains two versions—a basic version designed to support real-time implementations, and an advanced version optimized for the highest accuracy but not necessarily implementable in real-time. The primary applications for PEAQ are summarized in Table 2.6.

Table 2.6 Target Applications for ITU-R Rec. BS.1116 PEAQ

Category	Application	Version
Diagnostic	Assessment of implementations	Both
	Equipment or connection status	Advanced
	Codec identification	Both
Operational	Perceptual quality line-up	Basic
	On-line monitoring	Basic
Development	Codec development	Both
	Network planning	Both
	Aid to subjective assessment	Advanced

2.8.10 Perspective on Audio Compression

A balance must be struck between the degree of compression available and the level of distortion that can be tolerated, whether the result of a single coding pass or the result of a number of passes, as would be experienced in a complex audio chain or network [15]. There have been many outstanding successes for digital audio data compression in communications and storage, and as long as the limitations of the various compression systems are fully understood, successful implementations will continue to grow in number.

Compression is a tradeoff and in the end you get what you pay for. Quality must be measured against the coding algorithm being used, the compression ratio, bit rate, and coding delay resulting from the process.

There is continued progress in expanding the arithmetical capabilities of digital signal processors, and the supporting hardware developments would seem to be following a parallel course. It is possible to obtain a single chip containing both encoder and decoder elements, including stereo capabilities. In every five year period, it is not unreasonable to expect a tenfold increase in the processing capabilities of a single DSP chip, thus, increasing flexibility and processing power. Speculation could point to an eventual position when a completely lossless algorithm with an extremely high compression ratio would become available. In any event, the art of compressing audio data streams into narrower and narrower digital pipes will undoubtedly continue.

2.9   References

1. Lakhani, Gopal: "Video Compression Techniques and Standards," *The Electronics Handbook*, Jerry C. Whitaker (ed.), CRC Press, Boca Raton, Fla., pp. 1273–1282, 1996.

2. Solari, Steve. J.: *Digital Video and Audio Compression*, McGraw-Hill, New York, 1997.

3. Netravali, A. N., and B. G. Haskell: *Digital Pictures, Representation, and Compression*, Plenum Press, 1988.

4. Gilge, M.: "Region-Oriented Transform Coding in Picture Communication," *VDI-Verlag, Advancement Report, Series 10*, 1990.

5. DeWith, P. H. N.: "Motion-Adaptive Intraframe Transform Coding of Video Signals," *Philips J. Res.*, vol. 44, pp. 345–364, 1989.

6. Isnardi, M., and T. Smith: "MPEG Tutorial," *Proceedings of the Advanced Television Summit*, Intertec Publishing, Overland Park, Kan., 1996.

7. Nelson, Lee J.: "Video Compression," *Broadcast Engineering*, Intertec Publishing, Overland Park, Kan., p. 42, October 1995.

8. Arvind, R., et al.: "Images and Video Coding Standards," *AT&T Technical J.*, p. 86, 1993.

9. SMPTE 308M, "MPEG-2 4:2:2 Profile at High Level," SMPTE, White Plains, N.Y., 1998.

10. ATSC, "Guide to the Use of the ATSC Digital Television Standard," Advanced Television Systems Committee, Washington, D.C., doc. A/54, Oct. 4, 1995.

11. "IEEE Standard Specifications for the Implementation of 8×8 Inverse Discrete Cosine Transform," std. 1180-1990, Dec. 6, 1990.

12. Nelson, Lee J.: "Video Compression," *Broadcast Engineering*, Intertec Publishing, Overland Park, Kan., pp. 42–46, October 1995.

13. Smith, Terry: "MPEG-2 Systems: A Tutorial Overview," Transition to Digital Conference, *Broadcast Engineering*, Overland Park, Kan., Nov. 21, 1996.

14. SMPTE Recommended Practice: RP 202 (Proposed), "Video Alignment for MPEG-2 Coding," SMPTE, White Plains, N.Y., 1999.

15. Wylie, Fred: "Audio Compression Technologies," in *NAB Engineering Handbook*, 9th ed., Jerry C. Whitaker (ed.), National Association of Broadcasters, Washington, D.C., 1998.

16. Wylie, Fred: "Audio Compression Techniques," *The Electronics Handbook*, Jerry C. Whitaker (ed.), CRC Press, Boca Raton, Fla., pp. 1260–1272, 1996.

17. Lyman, Stephen, "A Multichannel Audio Infrastructure Based on Dolby E Coding," *Proceedings of the NAB Broadcast Engineering Conference*, National Association of Broadcasters, Washington, D.C., 1999.

18. Terry, K. B., and S. B. Lyman: "Dolby E—A New Audio Distribution Format for Digital Broadcast Applications," *International Broadcasting Convention Proceedings*, IBC, London, England, pp. 204–209, September 1999.

2.10 Bibliography

Bennett, Christopher: "Three MPEG Myths," *Proceedings of the 1996 NAB Broadcast Engineering Conference*, National Association of Broadcasters, Washington, D.C., pp. 129–136, 1996.

Bonomi, Mauro: "The Art and Science of Digital Video Compression," *NAB Broadcast Engineering Conference Proceedings*, National Association of Broadcasters, Washington, D.C., pp. 7–14, 1995.

Brandenburg, K., and Gerhard Stoll: "ISO-MPEG-1 Audio: A Generic Standard for Coding of High Quality Digital Audio," *92nd AES Convention Proceedings*, Audio Engineering Society, New York, N.Y., 1992, revised 1994.

Dare, Peter: "The Future of Networking," *Broadcast Engineering*, Intertec Publishing, Overland Park, Kan., p. 36, April 1996.

Fibush, David K.: "Testing MPEG-Compressed Signals," *Broadcast Engineering*, Overland Park, Kan., pp. 76–86, February 1996.

Freed, Ken: "Video Compression," *Broadcast Engineering*, Overland Park, Kan., pp. 46–77, January 1997.

IEEE Standard Dictionary of Electrical and Electronics Terms, ANSI/IEEE Standard 100-1984, Institute of Electrical and Electronics Engineers, New York, 1984.

Jones, Ken: "The Television LAN," *Proceedings of the 1995 NAB Engineering Conference*, National Association of Broadcasters, Washington, D.C., p. 168, April 1995.

Smyth, Stephen: "Digital Audio Data Compression," *Broadcast Engineering*, Intertec Publishing, Overland Park, Kan., February 1992.

Stallings, William: *ISDN and Broadband ISDN*, 2nd Ed., MacMillan, New York.

Taylor, P.: "Broadcast Quality and Compression," *Broadcast Engineering*, Intertec Publishing, Overland Park, Kan., p. 46, October 1995.

Whitaker, Jerry C., and Harold Winard (eds.): *The Information Age Dictionary*, Intertec Publishing/Bellcore, Overland Park, Kan., 1992.

Chapter 3

Program and System Information Protocol

Jerry C. Whitaker, Editor

3.1   Introduction

The *program and system information protocol* (PSIP) is a collection of tables designed to operate within every transport stream for terrestrial broadcast of digital television [1]. The purpose of the protocol, described in ATSC document A/65, is to specify the information at the system and event levels for all virtual channels carried in a particular transport stream. Additionally, information for analog channels—as well as digital channels from other transport streams—may be incorporated.

The typical 6 MHz channel used for analog broadcast supports about 19 Mbits/s throughput. Because program signals with standard resolution can be compressed using MPEG-2 to sustainable data rates of approximately 6 Mbits/s, three or four standard-definition (SD) digital television channels can be safely supported within a single physical channel. Moreover, enough bandwidth remains within the same transport stream to provide several additional low bandwidth nonconventional services such as:

- Weather reports
- Stock reports
- Headline news
- Software download (for games or enhanced applications)
- Image-driven classified ads
- Home shopping
- Pay-per-view information

It is, therefore, practical to anticipate that in the future, the list of services (*virtual channels*) carried in a physical transmission channel (6 MHz of bandwidth for the U.S.) may easily reach ten or more. Furthermore, the number and types of services may also change continuously, thus becoming a dynamic medium for entertainment, information, and commerce.

An important feature of terrestrial broadcasting is that sources follow a distributed information model rather than a centralized one. Unlike cable or satellite, terrestrial service providers are geographically distributed and have no interaction with respect to data unification or even synchronization. It is, therefore, necessary to develop a protocol for describing *system information* and *event descriptions* that are followed by every organization in charge of a physical transmission channel. System information allows navigation of and access to each of the channels within the transport stream, whereas event descriptions give the user content information for browsing and selection.

3.2   Elements of PSIP

PSIP is a collection of hierarchically associated tables, each of which describes particular elements of typical digital television services [1]. Figure 3.1 shows the primary components and the notation used to describe them. The packets of the base tables are all labeled with a *base PID* (base_PID). The base tables are:

- *System time table* (STT)
- *Rating region table* (RRT)
- *Master guide table* (MGT)
- *Virtual channel table* (VCT)

The *event information tables* (EIT) are a second set of tables whose packet identifiers are defined in the MGT. The *extended text tables* (ETT) are a third set of tables, and similarly, their PIDs are defined in the MGT.

The system time table is a small data structure that fits in one packet and serves as a reference for time-of-day functions. Receivers can use this table to manage various operations and scheduled events.

Transmission syntax for the U.S. voluntary program rating system is included in the ATSC standard. The rating region table has been designed to transmit the rating standard in use for each country using the system. Provisions also have been made for multicountry regions.

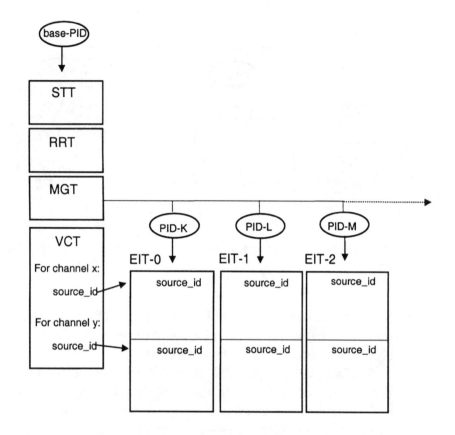

Figure 3.1 Overall structure for the PSIP tables. (*From* [1]. *Used with permission.*)

The master guide table provides general information about all of the other tables that comprise the PSIP standard. It defines table sizes necessary for memory allocation during decoding, defines version numbers to identify those tables that need to be updated, and generates the packet identifiers that label the tables.

The virtual channel table, also referred to as the *terrestrial VCT* (TVCT), contains a list of all the channels that are or will be on-line, plus their attributes. Among the attributes given are the channel name, navigation identifiers, and stream components and types.

As part of PSIP, there are several event information tables, each of which describes the events or television programs associated with each of the virtual channels listed in the VCT. Each EIT is valid for a time interval of 3

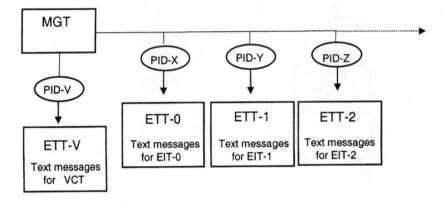

Figure 3.2 Extended text tables in the PSIP hierarchy. (*From* [1]. *Used with permission.*)

hours. Because the total number of EITs is 128, up to 16 days of programming may be advertised in advance. At minimum, the first four EITs must always be present in every transport stream.

As illustrated in Figure 3.2, there can be several extended text tables, each defined in the MGT. As the name implies, the purpose of the extended text table is to carry text messages. For example, for channels in the VCT, the messages can describe channel information, cost, coming attractions, and other related data. Similarly, for an event such as a movie listed in the EIT, the typical message would be a short paragraph that describes the movie itself. Extended text tables are optional in the ATSC system.

3.2.1 A/65 Technical Corrigendum and Amendment

During the summer of 1999, the ATSC membership approved a Corrigendum and an Amendment to the PSIP Standard (A/65). The *Technical Corrigendum No.1 to ATSC Standard A/65* documented changes for clarification, added an informative annex describing how PSIP could be used over cable, and made editorial corrections. The *Amendment No.1 to ATSC Standard A/65* made a technical change by defining a previously reserved bit to enable the unambiguous communication of virtual channel identification for channels that are not currently being used.

PSIP for Cable Applications

As outlined previously in this chapter, certain data specified in the Program and System Information Protocol (PSIP) forms a mandatory part of every ATSC-compliant digital multiplex signal delivered via terrestrial broadcast [2]. ATSC Document A/66, Annex G, provides an overview of the use of PSIP for digital cable applications.

PSIP was designed, as much as possible, to be independent of the physical system used to deliver the MPEG-2 multiplex. Therefore, the system time table, master guide table, virtual channel table (VCT), and event information tables and extended text tables are generally applicable equally as well to cable as to terrestrial broadcast delivery methods. The differences can be summarized as follows:

- For cable, the *cable virtual channel table* (CVCT) provides the VCT function, while the terrestrial virtual channel table applies for terrestrial broadcast. The cable VCT includes two parameters not applicable to the terrestrial broadcast case, and the syntax of several parameters in the table is slightly different for cable compared with the terrestrial broadcast case.

- Use of the program guide portion of PSIP (EIT and ETT) for cable is considered optional, while it is mandatory when PSIP is used for terrestrial broadcasting. Cable operators are free to *not* provide any program guide data at all if they so choose, or provide the data in a format other than PSIP if they do support an EPG.

While the syntax of the cable and terrestrial VCTs are nearly identical, the cable VCT has two parameters not present in the terrestrial VCT: a "path select" bit, and a bit that can indicate that a given virtual channel is transported *out-of-band* (OOB). Also, the semantics of the major and minor channel number fields and the source ID differ for the cable VCT compared with its terrestrial broadcast counterpart.

Channel Numbers

When PSIP is used for terrestrial broadcast, care must be taken in the assignment of major and minor channel numbers to avoid conflicts [2]. For example, the PSIP standard indicates that for the U.S. and its possessions, a terrestrial broadcaster with an existing NTSC license must use a major channel number for digital services that corresponds to the NTSC RF channel number in present use for the analog signal. For cable, such restrictions are technically unnecessary. For terrestrial broadcast, the major channel number is limited to the range 1 to 99 for ATSC digital television or audio services. For cable, major channel numbers can range from 1 to 999.

PSIP Data on Cable

PSIP data carried on cable in-band is analogous to PSIP included in the terrestrial digital broadcast multiplex: a receiver can discover the structure of digital services carried on that multiplex by collecting the current VCT from it [2]. A cable-ready digital TV can visit each digital signal on the cable, in sequence, and record from each a portion of the full cable VCT. This is exactly the same process a terrestrial digital broadcast receiver performs to build the terrestrial channel map.

Re-Multiplexing Issues

A cable operator may wish to take incoming digital transport streams from various sources (terrestrial broadcast, satellite, or locally generated), add or delete services or elementary streams, and then re-combine them into output transport streams [2]. If the incoming transport streams carry PSIP data, care must be taken to properly process this data in the re-multiplexer. Specifically, the re-multiplexer needs to account for any MPEG or PSIP fields or variables that are scoped to be unique within the transport stream. Such fields include PID values, MPEG program numbers, certain source ID tags, and event ID fields.

Enhancements to the PSIP Standard

In May 2000, the ATSC revised the PSIP standard to include an amendment that provides functionality known as *Directed Channel Change* (DCC), and also clarified existing aspects of the standard. The new feature allows broadcasters to tailor programming or advertising based upon viewer demographics. For example, viewers who enter location information such as their zip code into a DCC-equipped receiver can receive commercials that provide specific information about retail stores in their neighborhood. Segments of newscasts, such as weather reports can also be customized based upon this location information.

A channel change may also be based upon the subject matter of the content of the program. Nearly 140 categories of subject matter have been tabulated that can be assigned to describe the content of a program. A broadcaster can use this category of DCC request switching to direct a viewer to a program based upon the viewer's desire to receive content of that subject matter.

3.2.2 Acronyms and Abbreviations

The following acronyms and abbreviations are used within the ATSC PSIP specification [1]:

ATSC	Advanced Television Systems Committee
bslbf	bit serial, leftmost bit first
BMP	Basic Multilingual Plane
CAT	Conditional Access Table
CRC	Cyclic Redundancy Check
CVCT	Cable Virtual Channel Table
DTV	Digital Television
EPG	Electronic Program Guide
EIT	Event Information Table
EMM	Entitlement Management Message
ETM	Extended Text Message
ETT	Extended Text Table
GPS	Global Positioning System
PSIP	Program and System Information Protocol
MGT	Master Guide Table
MPAA	Motion Picture Association of America
MPEG	Moving Picture Experts Group
NVOD	Near Video On Demand
OOB	Out of Band
PAT	Program Association Table
PCR	Program Clock Reference
PES	Packetized Elementary Stream

PID	Packet Identifier
PMT	Program Map Table
PTC	Physical Transmission Channel
SCTE	Society of Cable Telecommunications Engineers
SI	System Information
STD	System Target Decoder
STT	System Time Table
rpchof	remainder polynomial coefficients, highest order first
RRT	Rating Region Table
TS	Transport Stream
TVCT	Terrestrial Virtual Channel Table
unicode	Unicode™
UTC	Coordinated Universal Time[1]
uimsbf	unsigned integer, most significant bit first
VCT	Virtual Channel Table. Used in reference to either TVCT or CVCT.

3.2.3 Definition of Terms

The following terms are used throughout the ATSC PSIP document [1]:

descriptor A data structure of the format: *descriptor_tag*, *descriptor_length*, and a variable amount of data. The tag and length fields are each 8 bits. The length specifies the length of data that begins immediately following the *descriptor_length* field itself. A descriptor whose *descriptor_tag* identifies a type not recognized by a particular decoder shall be ignored by that decoder. Descriptors can be included in certain specified places within PSIP tables, subject to certain restrictions. Descriptors may be used to extend data represented as fixed fields within the tables. They make the protocol very flexible because they can be included only as needed. New descriptor types can be standardized

---

1. Because unanimous agreement could not be achieved by the ITU on using either the English word order, CUT, or the French word order, TUC, a compromise to use neither was reached.

and included without affecting receivers that have not been designed to recognize and process the new types.

digital channel A set of one or more digital elementary streams. See *virtual channel*.

event A collection of elementary streams with a common time base, an associated start time, and an associated end time. An event is equivalent to the common industry usage of "television program."

instance See *table instance*.

logical channel See *virtual channel*.

physical channel A generic term to refer to the each of the 6-8 MHz frequency bands where television signals are embedded for transmission. Also known as the *physical transmission channel* (PTC). One analog virtual channel fits in one PTC but multiple digital virtual channels typically coexist in one PTC.

physical transmission channel See *physical channel*.

program element A generic term for one of the elementary streams or other data streams that may be included in a program. For example: audio, video, data, and so on.

program A collection of program elements. Program elements may be elementary streams. Program elements need not have any defined time base; those that do have a common time base are intended for synchronized presentation. The term *program* is also commonly used in the context of a "television program" such as a scheduled daily news broadcast. In this specification the term "event" is used to refer to a "television program" to avoid ambiguity.

region As used in the PSIP document, a region is a geographical area consisting of one or more countries.

section A data structure comprising a portion of an ISO/IEC 13818-1 defined table, such as the Program Association Table (PAT), Conditional Access Table (CAT), or Program Map Table (PMT). All sections begin with the *table_id* and end with the *CRC_32 field*, and their starting points within a packet payload are indicated by the *pointer_field* mechanism defined in the ISO/IEC 13818-1 International Standard.

stream An ordered series of bytes. The usual context for the term *stream* is the series of bytes extracted from transport atream packet payloads which have a common unique PID value (e.g., video PES packets or Program Map Table sections).

table PSIP is a collection of tables describing virtual channel attributes, event features, and others. PSIP tables are compliant with the private section syntax of ISO/IEC 13818-1.

table instance Tables are identified by the *table_id* field. However, in cases such as the RRT and EIT, several instances of a table may be defined simultaneously. All instances have the same PID and *table_id* but a different *table_id_extension*.

virtual channel A virtual channel is the designation, usually a number, that is recognized by the user as the single entity that will provide access to an analog TV program or a set of one or more digital elementary streams. It is called "virtual" because its identification (name and number) may be defined independently from its physical location. Examples of virtual channels include: digital radio (audio only), a typical analog TV channel, a typical digital TV channel (composed of one audio and one video stream), multi-visual digital channels (composed of several video streams and one or more audio tracks), or a data broadcast channel (composed of one or more data streams). In the case of an analog TV channel, the virtual channel designation will link to a specific physical transmission channel. In the case of a digital TV channel, the virtual channel designation will link both to the physical transmission channel and to the particular video and audio streams within that physical transmission channel.

3.2.4 Conditional Access System

One of the important capabilities of the DTV standard is support for delivering pay services through *conditional access* (CA) [3]. The ATSC, in document A/70, specified a CA system for terrestrial broadcast that defines a middle protocol layer in the ATSC DTV system. The standard does not describe precisely all the techniques and methods to provide CA, nor the physical interconnection between the CA device (typically a "smart card" of some type) and its host (the DTV receiver or set-top box). Instead, it provides the data envelopes and transport functions that allow several CA systems of different types to operate simultaneously. In other words, a broadcaster can offer pay-TV services by means of one or more CA systems, each of which may have different transaction mechanisms and different security strategies.

Such services generally fall into one of five major categories:

- *Periodic subscription*, where the subscriber purchases entitlements, typically valid for one month.

- *Order-ahead pay-per-view* (OPPV), where the subscriber pre-pays for a special event.

- *Pay-per-view* (PPV) and *impulse PPV* (IPPV), distinguished by the consumer deciding to pay close (or very close) to the time of occurrence of the event.

- *Near video on demand* (NVOD), where the subscriber purchases an event that is being transmitted with multiple start times. The subscriber is connected to the next showing, and may be able to pick up a later transmission of the same program after a specified pause.

All of these services can be implemented without requiring a return channel in the DTV system by storing a "balance" in the CA card and then "deducting" charges from that balance as programs are purchased. While video is implied in each of these services, the CA system can also be used for paid data delivery.

The ATSC standard uses the SimulCrypt technique that allows the delivery of one program to a number of different decoder populations that contain different CA systems and also for the transition between different CA systems in any decoder population. *Scrambling* is defined as a method of continuously changing the form of a data signal so that without suitable access rights and an electronic descrambling key, the signal is unintelligible. *Encryption*, on the other hand, is defined as a method of processing keys needed for descrambling, so that they can be conveyed to authorized users. The information elements that are exchanged to descramble the material are the *encrypted keys*. The ATSC key is a 168-bit long data element and while details about how it is used are secret, essential rules to enable coexistence of different CA systems in the same receiver are in the ATSC standard.

System Elements

The basic elements of a conditional access system for terrestrial DTV are the headend broadcast equipment, the conditional access resources, a DTV host,

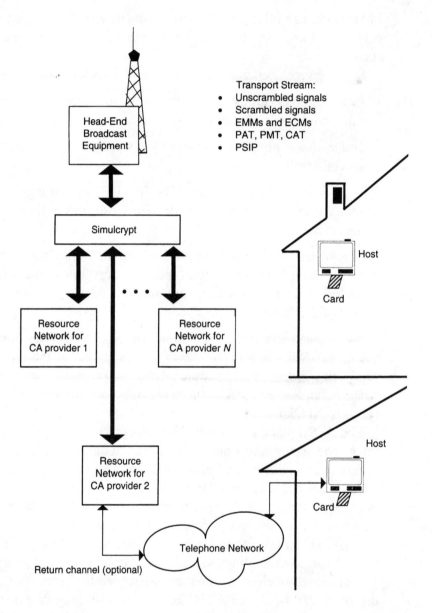

Transport Stream:
- Unscrambled signals
- Scrambled signals
- EMMs and ECMs
- PAT, PMT, CAT
- PSIP

Figure 3.3 The principle elements of a conditional access system as defined in ATSC A/70. (*After* 3].)

and the security module(s) [4]. Figure 3.3 illustrates these basic elements and some possible interactions among them. The headend broadcast equipment generates the scrambled programs for over-the-air transmission to the constellation of receivers. The DTV host demodulates the transmitted signals and passes the resulting transport stream to the security module for possible descrambling.

Security modules are distributed by CA providers in any of a number of ways. For example, either directly, through consumer electronics manufacturers, through broadcasters, or through their agents. Security modules typically contain information describing the status of the subscriber. Every time the security module receives from the host a TS with some of its program components scrambled, the security module will decide, based on its own information and information in the TS, if the subscriber is allowed access to one or more of those scrambled programs. When the subscriber is allowed access then the security module starts its most intensive task, the descrambling of the selected program.

The packets of the selected program are descrambled one by one in real time by the security module, and the resulting TS is passed back to the DTV host for decoding and display. According to the ATSC A/70 standard, two types of security module technologies are acceptable: NRSS-A and NRSS-B. A DTV host with conditional access support needs to include hardware and/or software to process either A or B, or both. The standard does not define a communication protocol between the host and the security module. Instead, it mandates the use of NRSS. Similarly, for copy protection of the interface between the host and the security module, the standard relies on NRSS specifications.

Besides the scrambled programs, the digital multiplex carries streams dedicated to the transport of *entitlement control messages* (ECM) and *entitlement management messages* (EMM). ECMs are data units that mainly carry the key for descrambling the signals. EMMs provide general information to subscribers and most likely contain information about the status of the subscription itself. Broadcasters interested in providing conditionally accessed services through one or more CA providers need to transmit ECMs and EMMs for each of those CA providers. The ATSC A/70 standard defines only the envelope for carrying ECMs and EMMs. A security module is capable of understanding the content of EMMs and ECMs privately defined by one (or more) CA provider.

The digital multiplex for terrestrial broadcast carries a program guide according to the specifications defined in ATSC document A/65. The program guide contains detailed information about present and future events that may be useful for the implementation of a CA system. For this reason, the A/70 standard defines a descriptor that can be placed in either the channel or event tables of A/65. Similar to the definitions of ECMs and EMMs, only the generic descriptor structure is defined while its content is private. The host is required to parse the program guide tables in search of this descriptor and pass it to the security module. Because of the private nature of the content, it is the security module that ultimately processes the information. Note that although A/65 PSIP does not require use of conditional access, the conditional access standard (A/70) requires the use of A/65 PSIP.

Most of the communications between the CA network and a subscriber receiver can be performed automatically through broadcast streams using EMMs. EMMs are likely to be addressed to a specific security module (or receiver, possibly groups of security modules or receivers). Security modules will receive their EMMs by monitoring the stream of EMMs in the multiplex, or by searching for addressed EMMs by *homing*. Homing is the method for searching EMM streams while the receiver is in stand-by mode. Homing is initiated by the host according to schedules and directives as defined in NRSS specifications for the security module. As Figure 3.3 shows, the security module can communicate with CA network using a telephone modem integrated into the host. This return channel is optional according to the A/70 standard, but if it exists, the host and the security module must adhere to the communication resource specifications of NRSS. A combination of homing and return channel EMM delivery can be used at the discretion of the CA provider.

Figure 3.3 shows that the interconnection between CA networks and the broadcast headend equipment requires Simulcrypt. Simulcrypt is a DVB protocol defining equipment and methods for adequate information exchange and synchronization. The most important information elements exchanged are the scrambling keys. According to Simulcrypt procedures, a new key can be generated by the headend equipment after a certain time interval that ranges from a fraction of a second to almost two hours. Before the encoder activates the new key, it is transferred to each CA system for encapsulation using their own protocols. Encapsulated keys in the form of ECMs are transferred back to the transmission equipment and are broadcast to all receivers. Shortly after the encoder has transmitted the new ECMs, the encoder uses the new key to scramble the content.

3.2.5 Transport Stream Identification

One of the many elements of configuration information needed to transmit a DTV bit stream that complies with the ATSC standard is a unique identification number [5]. The Transport Stream Identifier (TSID) was defined by the MPEG committee in ISO/IEC 13818-1, which is the Systems volume of the MPEG-2 standard. It also is required by ATSC Standard A/65 Program and System Information Protocol for Terrestrial Broadcast and Cable (PSIP).

The TSID is a critical number that provides DTV sets the ability to identify and tune a channel even if its frequency of operation is changed. It is a key link to the creation of *electronic program guides* (EPGs) to present the available program choices to the consumer. If the TSID is not a unique number assignment, then a receiver can fail to correctly associate all programs with the correct broadcaster or cable programmer. When present and unique, this value can not only enable EPGs, but can even help a DTV set find a broadcaster's DTV signal when it is carried on cable. When the DTV signal is put on a cable as an 8-VSB signal on a different RF channel (relative to the over-the-air broadcast), the TSID provides the link to the channel identity of the broadcaster. The TSID lets the receiver remap the RF frequency and select the program based on the major/minor channel number in the PSIP data stream. A DTV transport stream that is delivered via QAM retains the broadcaster's PSIP identification because the TSID in the cable system's Program Association Table (PAT) could be different. The TSID also permits use of DTV translators that appear to the consumer as if they had not been frequency shifted (i.e., they look like the main transmission, branded with the NTSC channel number).

To accomplish this functionality, the ATSC started with the ISO/IEC 13818-1 standard, which defined the TSID. It is a 16-bit number that must be placed in the PAT. According to MPEG-2, it serves as a label to identify the Transport Stream from any other multiplex within a network. MPEG-2 left selection of its value to the user.

As an element of the PSIP Standard (A/65), the ATSC has defined additional functions for the TSID and determined that its value for terrestrial broadcasts must be unique for a "network" consisting of geographically contiguous areas. The first such network is North America. The TSID is carried in the Virtual Channel Table (both cable and terrestrial versions) in PSIP.

The ATSC also established an identification number for existing analog television stations that is paired with the DTV TSID (differing in the least significant bit only). This Transmission Signal ID number is carried in the XDS packets associated with the closed captioning system on VBI line 21. The Consumer Electronics Manufacturers Association (CEMA) formalized how to carry this optional identifier in EIA 752. This number can provide a precise linkage between the NTSC service and the DTV service. Also, if the

NTSC channel is RF shifted and contains the complementary TSID, it then can be located by DTV sets and labeled with the original RF channel number.

In March 1998, the ATSC asked the FCC to assign and maintain TSID numbers for DTV broadcasters. At this writing, the FCC had not taken formal action on the request. When consumer DTV receivers were initially deployed, it was discovered that some models incorrectly identified channels because the broadcasters had not coordinated TSID assignments and, therefore, were transmitting the default TSID set by the multiplexer manufacturers (usually 0X0000 or 0X0001). The problem was brought to the attention of the Technical Committee of the HDTV Model Station Project, which then created a list of proposed TSIDs for U.S. DTV broadcasters. It was hoped that the FCC would use this list as the starting point for their maintenance of the assignments and coordination of assignments for Canada and Mexico.

3.3 Closed Captioning

On July 31, 2000, the FCC issued a Report and Order (R&O) in ET Docket No. 99-254 regarding DTV Closed Captions (DTVCC) [6]. The R&O amended Part 15 of the FCCs Rules, adopting technical standards for the display of closed captions on digital television receivers. It also amended Part 79 to require all captions to be passed through program distribution facilities and reflect the changes in Part 15. The R&O also clarified the compliance date for including closed captions in digital programming.

In 1990, Congress passed the Telecommunications Decoder Circuitry Act (TDCA), which required television receivers with picture screen diagonals of 13-in. or larger to contain built-in closed caption decoders and have the ability to display closed captioned television transmissions. The Act also required the FCC to take appropriate action to ensure that closed captioning services continue to be available to consumers as new technology was developed. In 1991, the FCC amended its rules to include standards for the display of closed captioned text on analog NTSC TV receivers. The FCC said that with the advent of DTV broadcasting, it would again update its rules to fulfill its obligations under the TDCA.

The R&O adopted Section 9 of EIA-708-B, which specifies the methods for encoding, delivery, and display of DTVCC. Section 9 recommends a minimum set of display and performance standards for DTVCC decoders. However, based on comments filed by numerous consumer advocacy groups, the FCC decided to require DTV receivers to support display features beyond those contained in Section 9. In addition, the FCC incorporated by reference the remaining sections of EIA-708-B into its rules for informational purposes only.

Manufacturers must begin to include DTVCC functionality, in accordance with the rules adopted in the R&O, in DTV devices manufactured as of July 1, 2002. Specifically:

- All digital television receivers with picture screens in the 4:3 aspect ratio measuring at least 13-in. diagonally.

- Digital television receivers with picture screens in the 16:9 aspect ratio measuring 7.8-in. or larger vertically (this size corresponds to the vertical height of an analog receiver with a 13-in. diagonal screen).

- All DTV tuners shipped in interstate commerce or manufactured in the U.S. The rules apply to DTV tuners whether or not they are marketed with display screens.

The R&O further stated that programming prepared or formatted for display on digital television receivers before the July 1, 2002, date that digital television decoders are required to be included in digital television devices is considered "pre-rule" programming (as defined in the FCC's existing the closed captioning rules). Therefore, programming prepared or formatted for display on digital television after that date will be considered *new programming*. The existing rules require an increasing amount of captioned new programming over an eight-year transition period with 100 percent of all new non-exempt programming required to be captioned by January 1, 2006. This means that as of July 1, 2002, DTV services have the same hourly captioning requirement as NTSC services. The average amount required per day in 2002 is nearly 10 hrs (900 hrs/quarter). Those stations operating for part of a quarter are expected to meet the prorated or average daily amount.

There are three ways that the stations can originate DTVCC:

- If the DTV captions arrive already formatted and embedded in an MPEG-2 video stream, then the broadcaster is required to insure they are passed through and transmitted to receivers.

- If the DTV program is being up-converted from an NTSC source, then the caption data in that NTSC program must, at a minimum, be encapsulated into EIA-708-B format captions (using CC types 00 or 01) and broadcast with the DTV program.

- If the program is locally originated (and not exempt) and captions are being locally created but are not in DTVCC. then again—at a minimum—the caption information must be encapsulated into EIA-708-B format.

In the R&O, the FCC also stated that in order for cable providers to meet their closed captioned obligations, they must pass through closed captions they receive to digital television sets. Also, they must transmit those captions in a format that will be understandable to DTVCC decoders.

Regarding DTV set-top boxes, converter boxes, and standalone tuners, if these devices have outputs that are intended to be used to display digital programming on analog receivers, then the device must deliver the encoded "analog" caption information on those outputs to the attached analog receiver.

3.3.1 SMPTE 333M

To facilitate the implementation of closed-captioning for DTV facilities, the SMPTE developed SMPTE 333M, which defines rules for interoperation of DTVCC data server devices and video encoders. The caption data server devices provide partially-formatted EIA 708-A data to the DTV video encoder using a *request/response* protocol and interface, as defined in the document. The video encoder completes the formatting and includes the EIA 708-A data in the video elementary stream.

3.4 Data Broadcasting

It has long been felt that data broadcasting will hold one of the keys to profitability for DTV stations, at least in the early years of implementation. The ATSC Specialist Group on Data Broadcasting, under the direction of the Technology Group on Distribution (T3), was charged with investigating the transport protocol alternatives to add data to the suite of ATSC digital television standards [7]. The Specialist Group subsequently prepared a standard to address issues relating to data broadcasting using the ATSC DTV system.

The foundation for data broadcasting is the same as for video, audio, and PSIP—the MPEG-2 standard for transport streams (ISO/IEC 13818-1). Related work includes the following:

- ISO standardization of the Digital Storage Media Command Control framework in ISO/IEC 13818-6.

- The Internet Engineering Task Force standardization of the Internet Protocol in RFC 791.

- The ATSC specification of the data download protocol, addressable section encapsulation, data streaming, and data piping protocols.

The service-specific areas and the applications are not standardized. The DTV data broadcasting standard, in conjunction with the other referenced standards, defines how data can be transported using four different methods:

- *Data piping*, which describes how to put data into MPEG-2 transport stream packets. This approach supports private data transfer methods to devices that have service-specific hardware and/or software.

- *Data streaming*, which provides additional functionality, especially related to timing issues. The standard is designed to support synchronous data broadcast, where the data is sent only once (much as the video or audio is sent once). The standard is based on PES packets as defined by MPEC-2.

- *Addressable section encapsulation*, built using the DSM-CC framework. The ATSC added specific information to customize the framework for the ATSC environment, especially in conjunction with the PSIP standard, while retaining maximum commonality with the DVB standards. These methods enable repeated transmission of the same data elements, thus enabling better availability or reliability of the data.

- Data download

Because receivers will have different capabilities as technology evolves, methods to enable the receivers to determine if they could support data services also were developed. This type of "data about the data" is referred to as *control data* or *metadata*. The control information describes where and when the data service is being transmitted and provides linkage information. This standard uses and builds upon the PSIP standard. A *data information table* (DIT), which is structured in a manner similar to an *event information table* (EIT), transmits the information for data-only services. For data that is closely related to audio or video, the EIT can contain announcement information as well. Each data service is announced with key information about its data rate profile and receiver buffering requirements. Both opportunistic and fixed data rate allocations are defined in four profiles. These data facilitate receivers only presenting services to the consumer that the receiver can actually deliver. Each data service has its own minor channel number, which must have a value greater than 100.

Because data services can be quite complex, and related data might be provided to the receiver via different paths than the broadcast channel, an additional structure to standardize the linkage methods was developed. The structure consists of two tables—the *data service table* (DST) and the *network resource table* (NRT)—that together are called the *service description framework* (SDF). The DST contains one of thirteen protocol encapsulations and the linkage information for related data that is in the same MPEG TS. The NRT contains information to associate data streams that are not in the TS.

3.4.1 PSIP and SDF for Data Broadcasting

The PSIP and SDF are integral elements of the data broadcasting system. These structures provide two main functions [8]:

- Announcement of the available services

- Detailed instructions for assembling all the components of the data services as they are delivered

Generally, data of any protocol type is divided and subdivided into packets before transmission. PES packets are up to 64 kB long; sections are up to 4 kB, and transport stream (TS) packets are 188 bytes. Protocol standards document the rules for this orderly subdivision (and reassembly). All information is transmitted in 188 byte TS packets. The receiver sees the packets, and by using the packet identifier (PID) in the header of each packet, routes each packet to the appropriate location within the receiver so that the information can be recovered by reversing the subdivision process.

A data service can optionally be announced in either an EIT or a DIT, in conjunction with additional entries in the virtual channel table (VCT). These tables use the MPEG section structure. Single data services associated with a program are announced in the EIT. The DIT is used to support direct announcement of data services that are associated with a video program, but start and stop at different times within that program. Like the EIT, there are four required DITs that are used for separately announced data services. DITs contain the start time and duration of each event, the data ID number, and a data broadcast descriptor. This descriptor contains information about the type of service profile, the necessary buffer sizes, and synchronization information.

There are three profiles for services that need constant or guaranteed delivery rates:

- G1—up to 384 kbits/s

- G2—up to 3.84 Mbits/s

- G3—up to the full 19.4 Mbits/s transport stream

For services that are delivered opportunistically, at up to the full transport data rate, the profile is called A1. Also present are other data that enable the receiver to determine if it has the capability to support a service being broadcast.

The VCT contains the virtual channel for each data service. The ATSC PSIP data service is intended to facilitate human interaction through a program guide that contains a linkage to the SDF, which provides the actual road map for reassembly of the fragmented information. The process of following this roadmap is known as *discovery and binding*. The SDF information is part of the bandwidth of the data service, not part of the broadcaster's overhead bandwidth (PSIP is in the overhead).

As mentioned previously, the SDF contains two distinct structures, the DST and the NRT (each use MPEG-2 private sections). The concepts for application discovery and binding rely upon standard mechanisms defined in ISO/IEC-13818-6 (MPEG-2 Digital Storage Media Command and Control, DSM-CC). The MPEG-2 transport stream packets conveying the DST and the NRT for each data service are referenced by the same PID, which is different from the PID value used for the SDF of any other data service. The DST must be sent at least once during the delivery of the data service. Key elements in the DST include the following:

- A descriptor defining signal compatibility and the requirements of an application

- The name of the application

- Method of data encapsulation

- List of author-created associations and application data structure parameters

Some services will not need the NRT because it contains information to link to data services outside the TS.

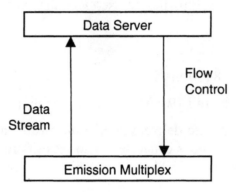

Figure 3.4 The opportunistic flow control environment of SMPTE 325M. (*After*[9].)

3.4.2 Opportunistic Data

The SMPTE 325M standard defines the flow control protocol to be used between an emission multiplexer and data server for *opportunistic data broadcast* [9]. Opportunistic data broadcast inserts data packets into the output multiplex to fill any available free bandwidth. The emission multiplexer maintains a buffer from which it draws data to be inserted. The multiplexer requests additional MPEG-2 transport packets from the data server as its buffer becomes depleted. The number of packets requested depends upon the implementation, with the most stringent requirement being the request of a single MPEG-2 transport packet where the request and delivery can occur in less than the emission time of an MPEG-2 transport packet from the multiplexer.

This protocol is designed to be extensible and provide the basis for low latency, real-time backchannel communications from the emission multiplexer. Encapsulated in MPEG-2 transport packets, the messages of the flow control protocol are transmitted via MPEG-2 DSM-CC sections, following the message format defined in ISO/IEC 13818-6, Chapter 2. Such sections provide the capability to support error correction or error detection (or to ignore either).

The environment for this standard is illustrated in Figure 3.4. Within an emission station, one or more data servers provide broadcast data (contained within MPEG-2 transport packets, with appropriate protocol encapsulations) to an emission multiplexer. A real-time control path is available from the emission multiplexer to the data server for flow control messages. Opportunistic data broadcast will attempt to fill any bandwidth available in the emission multiplex with broadcast data on a nearly instantaneous basis. The emission multiplexer is in control of the opportunistic broadcast because it is aware of the instantaneous gaps in the multiplex.

The operational model of SMPTE 325M is that the emission multiplexer will maintain an internal buffer from which it can draw MPEG-2 data packets to insert into the emission multiplex as opportunity permits. As the buffer is emptied, the mux requests a number of packets from the data server to maintain buffer fullness over the control path. These data packets are delivered over the data path. To avoid buffer overflow problems (should the data server be delayed in servicing the packet request), the following conventions are recommended:

- The data server should not queue requests for a given service (that is, a new request will displace one that has not been acted upon).

- The emission multiplexer should request no more than half of its buffer size at a time.

Support for multiple opportunistic streams (multiple data servers and multiple opportunistic broadcasts from a single server) is provided by utilizing the MPEG-2 transport header PID as a session identifier.

3.4.3 RP 203

SMPTE Recommended Practice 203 (proposed at this writing) defines the means of implementing opportunistic data flow control in a DTV MPEG-2 transport broadcast according to flow control messages defined in SMPTE 325M [10]. An emissions multiplexer requests opportunistic data packets as the need for them arises and a data server responds by forwarding data already inserted into MPEG-2 transport stream packets. The control protocol that allows this transfer of asynchronous data is extensible in a backward-compatible manner to allow for more advanced control as may be necessary in the future. Control messages are transmitted over a dedicated data link of sufficient quality to ensure reliable real-time interaction between the multiplexer and the data server.

3.4.4 ATSC A/90 Standard

With the foregoing efforts as a backdrop, the ATSC released document A/90 to define in specific terms data broadcast features, functions, and formats.

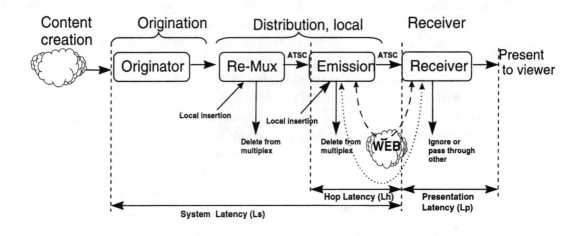

Figure 3.5 ATSC data broadcast system diagram. (*After* [11].)

The ATSC data broadcast system is illustrated in diagram form in Figure 3.5. The standard covers the delivery of data from the last part of the distribution chain (emission transmitter) to a receiver. While they are significant, issues related to delivery of data from originating points to this "last transmitter" using transport mechanisms other than ATSC transmission (such as disks, tapes, various types of network connections, and so on) are not described in the document.

Receivers are assumed to vary greatly in the number of services they are capable of presenting and their ability to store data or process it in some meaningful way [11]. Some may decode and present several audio/video broadcasts along with multiple data services. Others may be designed to perform a single function (such as delivering a stock ticker) as inexpensively as possible.

The A/90 standard defines the carriage of data using the *non-flow controlled scenario* and the *data carousel scenario* of the DSM-CC user-to-network download protocol. The ATSC use of the DSM-CC download protocol supports the transmission of the following:

- Asynchronous data modules

- Asynchronous data streaming

- Non-streaming synchronized data

Data carried by the download protocol may be error protected, since the DSM-CC sections used include a checksum field for that purpose.

The standard defines transmission of *datagrams* in the payload of MPEG-2 TS packets by encapsulating the datagrams in DSM-CC addressable sections. This mechanism is used for the asynchronous delivery of datagrams having the following characteristics:

- No MPEG-2 systems timing is associated with the delivery of data

- The smoothing buffer can go empty for indeterminate periods of time

- The data is carried in DSM-CC sections or DSM-CC addressable sections

The A/90 standard supports synchronous and synchronized data streaming using PES. *Synchronous data streaming* is defined as the streaming of data with timing requirements in the sense that the data and clock can be regenerated at the receiver into a synchronous data stream. Synchronous data streams have no strong timing association with other data streams and are carried in PES packets.

Synchronized data streaming implies a strong timing association between PES streams referenced by different PIDs. Synchronized streaming data are carried in PES packets. An example is application data associated with a video stream.

The standard defines *data piping* as a mechanism for delivery of arbitrary user defined data inside an MPEG-2 TS. Data are inserted directly into the payload of MPEG-2 TS packets. No methods are specified in the standard for fragmentation or re-assembly of data sent in this manner.

A *data service*, as defined in document A/90, is a collection of one or more data broadcast types. For example, a data service may include streaming synchronized data and asynchronous multiprotocol encapsulated data.

The tables, codes, and commands necessary for making this system work are described in [11].

3.4.5 Acronyms and Abbreviations

The following acronyms and abbreviations are used within the ATSC Data Broadcast Standard [11]:

ATSC   Advanced Television Systems Committee

bslbf   bit serial, leftmost bit first

CRC   Cyclic Redundancy Check

CVCT	Cable Virtual Channel Table
DASE	DTV Applications Software Environment
DAU	Data Access Unit
DEBn	Data Elementary Stream Buffer for synchronized data elementary stream n
DEBSn	Data Elementary Stream Buffer Size for synchronized data elementary stream n
DES	Data Elementary Stream
DET	Data Event Table
DSM-CC	Digital Storage Media Command and Control
DTS	Decoding Time-Stamp
DST	Data Service Table
DTV	Digital Television
DVB	Digital Video Broadcast
EIT	Event Information Table
ES	Elementary Stream
ETM	Extended Text Message
ETT	Extended Text Table
HTML	Hypertext Markup Language
IEC	International Electrotechnical Commission
IEEE	Institute of Electrical and Electronics Engineers
ISO	International Organization for Standardization
ITU	International Telecommunication Union
LLC-SNAP	Logical Link Control – Sub Network Access Protocol
MAC	Media Access Control
MGT	Master Guide Table
MPEG	Moving Picture Experts Group
MTU	Maximum Transmission Unit

NRT	Network Resources Table
OUI	Organization Unique Identifier
PAT	Program Association Table
PCR	Program Clock Reference
PES	Packetized Elementary Stream
PID	Packet Identifier
PMT	Program Map Table
PSI	Program Specific Information
PSIP	Program and System Information Protocol
PTS	Presentation Time Stamp
PU	Presentation Unit
rpchof	remainder polynomial coefficients, highest order first
RRT	Rating Region Table
SDT	Service Description Table
SCTE	Society of Cable Telecommunications Engineers
SI	System Information
STD	System Target Decoder
STT	System Time Table
TBn	Transport Buffer for data elementary stream n
TBSn	Transport Buffer Size for data elementary stream n
TCP/IP	Transmission Control Protocol/Internet Protocol
TS	Transport Stream
TVCT	Terrestrial Virtual Channel Table
UTC	Coordinated Universal Time[1]
uimsbf	Unsigned Integer, Most Significant Bit First

---

1. Because agreement could not be achieved by the ITU on using either the English word order, CUT, or the French word order, TUC, the compromise was to use neither.

nbomsbf Network Byte Order (most significant byte first), Most Significant Bit First.

VCT Virtual Channel Table

3.4.6 Global Terms

The following terms are used throughout the ATSC A/90 document [11]:

application An aggregation of related data items, including but not limited to procedural code, declarative data, and other data.

asynchronous data Stand-alone or audio/video-related data transmitted with no strong timing requirements in the sense that it is not associated with any transmitted clock references and availability of data in a data receiver is not governed by any such clock references.

audio-visual event An event (see definition below) where elementary streams are all of type video or audio.

ATSC Advanced Television Systems Committee. The committee responsible for the coordination and development of voluntary technical standards for advanced television systems.

bit rate The rate at which the bit stream is delivered from the channel to the input of a decoder.

bps Bits per second.

byte-aligned A bit in a coded bit stream is byte-aligned if its position is a multiple of 8-bits from the first bit in the stream.

communication channel A digital medium that transports a digital stream. A communication channel can be uni-directional or bi-directional.

constant bit rate Operation where the bit rate is constant from start to finish of the bit stream.

CRC The cyclic redundancy check used to verify the correctness of the data.

data access unit The portion of a synchronized or synchronous Data Elementary Stream that is associated with a particular MPEG-2 Presentation Time Stamp.

data carousel The scenario of the DSM-CC User-to-Network Download protocol that embodies the cyclic transmission of data.

data elementary stream The payloads of a series of consecutive MPEG-2 transport stream packets referenced by a unique PID value.

data element A self-contained subset of a data elementary stream.

data module An ordered sequence of bytes of a bounded size.

data receiver Any device capable of receiving and consuming data carried on an MPEG-2 Transport Stream.

data service A collection of applications and associated data elementary streams as signaled in a Data Service Table of the Service Description Framework. A data service is characterized by a profile and a level.

datagram A datagram is the fundamental protocol data unit in a packet-oriented data delivery protocol. Typically, a datagram is divided into header and data areas, where the header contains full addressing information (source and destination addresses) with each data unit. Datagrams are most often associated with connectionless network and transport layer services.

data source The provider of data that is being inserted into the MPEG-2 Transport Stream.

decoded stream The decoded reconstruction of a compressed bit stream.

decoder An embodiment of a decoding process.

decoding (process) The process defined in the Digital Television Standard that reads an input coded bit stream and outputs decoded pictures, audio samples, or data objects.

encoding (process) A process that reads a stream of input pictures or audio samples and produces a valid coded bit stream as defined in the Digital Television Standard.

event A collection of elementary streams with a common time base, an associated start time, and an associated end time. An event is equivalent to the common industry usage of "TV program."

forbidden This term, when used in clauses defining the coded bit stream, indicates that the value shall never be used. This is usually to avoid emulation of start codes.

Huffman coding A type of source coding that uses codes of different lengths to represent symbols which have unequal likelihood of occurrence.

instance See *table instance*.

Kbps 1,000 bits per second.

latency The total time from when a data object is transmitted in an MPEG-2 transport stream until the time it is fully decoded in the data receiver.

layer One of the levels in the data hierarchy of the video and system specification.

level The abstracted dimension that is used to refer to the size of the Data Elementary Buffer in the Transport System Target Decoder governing the delivery of Data Access Units of a Data Service.

logical channel See *virtual channel*.

Maximum Transmission Unit The largest amount of data that can be transferred in a single unit across a specific physical connection. When using the Internet Protocol, this translates to the largest IP datagram size allowed.

Mbps 1,000,000 bits per second.

MPEG Refers to standards developed by the ISO/IEC JTC1/SC29 WG11, *Moving Picture Experts Group*. MPEG may also refer to the Group itself.

MPEG-2 Refers to the collection of ISO/IEC standards 13818-1 through 13818-6.

multiplexer (Mux) A physical device that is capable of inserting MPEG-2 transport stream packets into and extracting MPEG-2 transport stream packets from an MPEG-2 transport stream.

multiprotocol encapsulation The encapsulation of datagrams in addressable sections.

opportunistic data Data inserted into the remaining available bandwidth in a given transport stream after all necessary bits have been allocated for video, audio, and other services.

packet A packet is a set of contiguous bytes consisting of a header followed by its payload.

packet identifier (PID) A unique integer value used to associate elementary streams of a program in a single or multi-program transport stream.

payload Payload refers to the bytes following the header byte in a packet.

PES packet header The leading fields in a PES packet up to but not including the *PES_packet_data_byte* fields where the stream is not a *padding stream*. In the case of a padding stream, the PES packet header is defined as the leading fields in a PES packet up to but not including the *padding_byte* fields.

PES packet The data structure used to carry elementary stream data. It consists of a packet header followed by PES packet payload.

PES stream A continuous sequence of PES packets of one elementary stream with one *stream_id*.

physical channel A generic term to refer to the each of the 6 to 8 MHz frequency bands where television signals are embedded for transmission. Also known as the *physical transmission channel* (PTC). One analog virtual channel fits in one PTC but multiple digital virtual channels typically coexist in one PTC. The calculations in this document are generally based on the ATSC 6 MHz channel capacity.

physical transmission channel See *physical channel*.

presentation time-stamp (PTS) A field that may be present in a PES packet header that indicates the time that a presentation unit is presented in the system target decoder.

presentation unit (PU) A decoded audio access unit or a decoded picture.

program A collection of program elements. Program elements may be elementary streams. Program elements need not have any defined time base; those that do have a common time base and are intended for synchronized presentation. The term *program* is also used in the context of a "television program" such as a scheduled daily news broadcast. In this Standard the term "event" is used for the latter to avoid ambiguity.

program clock reference (PCR) A time stamp in the transport stream from which decoder timing is derived.

program element A generic term for one of the elementary streams or other data streams that may be included in a program. For example: audio, video, data, and so on.

program specific information (PSI) PSI consists of normative data which is necessary for the demultiplexing of transport streams and the successful regeneration of programs.

profile A defined subset of data delivery characteristics.

PSIP Program and System Information Protocol is a collection of tables describing virtual channel attributes, event features, and other information.

reserved This term, when used in clauses defining the coded bit stream, indicates that the field may be used in the future for Digital Television Standard extensions.

scrambling The alteration of the characteristics of a video, audio, or coded data stream in order to prevent unauthorized reception of the information in a clear form. This alteration is a specified process under the control of a conditional access system.

section A data structure comprising a portion of an ISO/IEC 13818-1 or ISO/IEC 13818-6 defined table, such as the Program Association Table (PAT), Conditional Access Table (CAT), Program Map Table (PMT) or DSM-CC section. All sections begin with the *table_id* and end with the *CRC_32* or a checksum field, and their starting points within a packet payload are indicated by the *pointer_field* mechanism.

Service Description Framework The information conveyed in the program element and providing the Data Service Table and optionally the Network Resource Table of a single data service.

start codes 32-bit codes embedded in the coded bit stream that are unique. They are used for several purposes including identifying some of the layers in the coding syntax. Start codes consist of a 24-bit prefix (0x000001) and an 8-bit *stream_id*.

STD input buffer A first-in, first-out buffer at the input of a system target decoder for storage of compressed data from elementary streams before decoding.

stream An ordered series of bytes. The usual context for the term *stream* is the series of bytes extracted from transport stream packet payloads that have a common unique PID value (e.g., video PES packets or Program Map Table sections).

stream data Stream is a data object that has no specific start or end. The decoding system may need only a small fraction of the total data to activate a given application. An example includes stock ticker services.

synchronous data Data that uses MPEG-2 PCRs and MPEG-2 PTSs with the objective of delivering data units with timing constraints, these data units being processed for presentation and/or display as a standalone stream.

synchronized data Data that uses MPEG-2 PCRs and MPEG-2 PTSs with the objective of matching presentation and/or display of data units with access units of other streams (typically audio and video).

system target decoder (STD) A hypothetical reference model of a decoding process used to describe the semantics of the Digital Television Standard multiplexed bit stream.

table The collection of re-assembled sections bearing a common version number.

table instance Tables are identified by the *table_id* field. However, in cases such as the Data Event Table, several instances of a table are defined simultaneously. All instances are conveyed in transport stream packets of the same PID value and have the same *table_id* field value. Each instance has a different *table_id_extension* value.

Tap A reference to a data resource, including but not limited to a data elementary stream, a data carousel module, or a network resource.

time-stamp A term that indicates the time of a specific action such as the arrival of a byte or the presentation of a presentation unit.

transport stream Refers to the MPEG-2 transport stream syntax for the packetization and multiplexing of video, audio, and data signals for digital broadcast systems.

transport stream packet header The leading fields in a transport stream packet up to and including the *continuity_counter* field.

virtual channel A virtual channel is the designation, usually a number, that is recognized by the user as the single entity that will provide access to an analog TV program or a set of one or more digital elementary streams. It is called "virtual" because its identification (name and number) may be defined independently from its physical location.

3.4.7 Data Implementation Working Group

Because data broadcast is likely to become an important component of the emerging digital television broadcast services, the ATSC established a group to study implementation issues and make recommendations [12]. The *Data Implementation Working Group* (DIWG) recommendations were designed to foster interoperability between components. The scope of these recommendations includes the interconnections between functional data broadcast components within the emission station, management and provisioning parameters common to data broadcast equipment, flow control for opportunistic data broadcast, and tightly synchronized data.

DIWG has utilized a few simple guidelines in its activities:

• Do not reinvent the wheel—if a reasonable solution exists that can easily be modified to resolve the problem, recommend its use.

• Use a "minimal set" approach—where ever possible, recommend reuse of existing technology.

• Do not require unnatural functionality of existing equipment—aim for a near-term developable solution.

Environment

Figure 3.6 illustrates the logical components that comprise the model emission station environment from which DIWG has based its recommendations [12]. In practice, these functions can be physically grouped in a different manner than shown in the diagram. In addition, only certain classes of data have been considered in these recommendations, primarily data that can be viewed as content. Specifically, closed caption data has not been addressed.

Specific processes that comprise the DIWG's Model Emission Station environment include the following:

• The video and audio encoders read streams of video/audio samples and produce a coded bitstream as defined by ATSC A/53U.

• The program multiplexor (PMUX) combines a number of elementary streams (typically audio and video) into a single program. The output of the program multiplexor is an MPEG-2 transport stream.

• The conditional access generator is an optional entity, which is present only if some portion of the emission multiplexor output transport stream is to be encrypted.

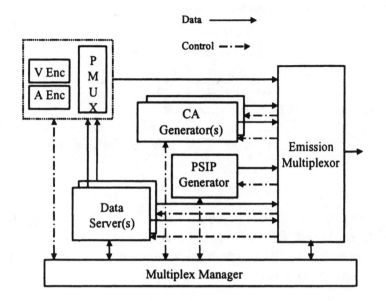

Figure 3.6 Functional diagram of the emission station. (*After* [12].)

- The PSIP generator provides the program and system information necessary for the multiplexor to generate an ATSC compliant MPEG-2 multi-program transport stream.

- The data server provides a valid coded representation using the protocols and techniques defined in the ATSC Data Broadcast Standard for any input data. The output of the data server is always correctly formatted MPEG-2 transport stream packets. The data server may also support the opportunistic data delivery protocol.

- The emission multiplexor is capable of inserting MPEG-2 transport stream packets into and extracting MPEG-2 transport stream packets from one or more MPEG-2 transport streams. The output of the emission multiplexor is a multi-program MPEG-2 transport stream. All input data except from the multiplex manager is of the form of MPEG-2 transport stream packets.

- The multiplex manager is controlled by a higher level entity that is outside the scope of the DIWG committee. The multiplex manager is responsible for communicating any required information to any entity needing said information.

Table 3.1 General Data Management Parameters (*After* [12].)

Data Server	CBR: PIDs, rates, handles
	Opportunistic: PIDs, flow control port configuration
	Output port configuration: connection type specific parameters
	Time for parameters to take effect
Emissions Multiplexor	CBR: PID remapping, data rates
	Opportunistic: PID remapping, flow control parameters, flow control port configuration
	Input port configuration: connection type specific parameters

In seeking to define interconnections between the components of a DTV data broadcast, DIWG sought to use readily available building blocks that preserve the quality of service requirements of a real-time broadcaster and leverage existing hardware and software.

Management and Control

Management of data broadcast equipment in the emission station is necessary in a similar manner as management of the video and audio equipment [12]. Configuration parameters must be communicated from the management system to the data server and emission multiplexor to allow data broadcast services to be delivered. Table 3.1 provides a short summary of the types of necessary management parameters for data broadcast. Parameters already managed for non-data services are not listed. DIWG has not recommended the mechanism for management, instead leaving this recommendation to one of the other groups working in this area (specifically, SMPTE).

Synchronization between elements of data broadcast and other content will likely become a requirement for many forms of television enhancement. Synchronization can be either *loose* (synchronization within a few seconds) or *tight* (synchronization at the field/frame level in the case of video). Typically, loose synchronization simply depends upon the presence of data or the reception of a message and does not require time stamps encapsulated within the data itself. This type of synchronization may be handled through the existing traffic or scheduling system. Tightly synchronized data, however, will require the presence of time stamps and careful control of emission and decode timing.

Synchronized data is defined as having a tight timing relationship to another program element. In this case, "tight timing relationship" means presentation accuracy within one video field. "Another program element" typically refers to a video elementary stream within the same MPEG program, but may refer to an audio elementary stream or even another data stream. It is important that the synchronized data stream reference the same MPEG timeline as the "synchronized to" stream (referring to the same Program Clock Reference (PCR) timeline). This typically means that the data and other streams are contained within the same MPEG program.

In the most common case, synchronized data refers to the presentation of visual information (coming from the data broadcast) at a prespecified video field (or frame). The intended timing relationship would be established during the authoring process and be conveyed to the ultimate consumer of the data. Difficulties arise because data and program are not tightly bound during the distribution process (in fact, they may take totally different paths from creation to emission). In addition, generic data is not bounded in complexity and the receiver is not bounded in simplicity, thus the decoding time for data at the receiver is not constrained.

Currently, the only timing information within the received MPEG transport stream is derived from the transmitted PCR fields. The PCR values are used to reconstruct a System Time Clock (STC), which is accurately synchronized to the encoder's clock. Video and audio information are presented to the viewer when their PTSs match the running STC. MPEG does not define any form of synchronization mechanism for generic data, but the model for video and audio may be used as a starting point. The PCR timeline does not necessarily monotonically increase. The PCR is carried within a finite-sized field (33 bits) and must roll over when the end is reached. Of more importance, PCR discontinuities are allowed, for example, when switching between encoders (running on different PCR timelines) for commercials.

DIWG developed the concept of Data Decode Delay time (DDD) which is author-specified as the amount of time between receipt of the final bit of data and its presentation time (this delay is used to decode the data into a form suitable for presentation). This DDD value may be quite large (seconds or even minutes) compared to the audio and video delays (which are typically subsecond). Because of the DDD, data may need to be sent early and precached, so that it may be decoded in time for presentation. In effect, the DDD value defines minimum decoding requirements in data receivers.

The MPEG buffer model allows data from at most two PCR timelines to coexist in the elementary buffer. In the case of complex data (long decode delay) combined with short commercials (each having its own PCR timeline), there may be data from more than two PCR timelines (which would violate the MPEG buffer model), requiring a special solution. In practice, this special case would likely be encountered when complex data is intended to be presented at the first frames or subseconds of a spliced commercial.

The synchronization problem is divided into two main cases. The first is when the PCR discontinuities are known and understood by the author, and the second is when the PCR discontinuities are either unknown or not understood by the author. The former case occurs for example when the author is colocated with the station where the data is finally embedded into the broadcast. The latter case will be likely to occur when the author is removed (physically or virtually) from the operation of the data combination station and/or is authoring content for multiple transport types.

3.5 References

1. ATSC: "Program and System Information Protocol for Terrestrial Broadcast and Cable," Advanced Television Systems Committee, Washington, D.C., Doc. A/65, February 1998.

2. ATSC: "Technical Corrigendum No.1 to ATSC Standard: Program and System Information Protocol for Terrestrial Broadcast and Cable," Doc. A/66, ATSC, Washington, D.C., December 17, 1999.

3. NAB: "Pay TV Services for DTV," *NAB TV TechCheck*, National Association of Broadcasters, Washington, D.C., October 4, 1999.

4. ATSC: "Conditional Access System for Terrestrial Broadcast," Advanced Television Systems Committee, Washington, D.C., Doc. A/70, July 1999.

5. *NAB TV TechCheck*, National Association of Broadcasters, Washington, D.C., January 4, 1999.

6. NAB: "Digital TV Closed Captions," *NAB TV TechCheck*, National Association of Broadcasters, Washington, D.C., August 7, 2000.

7. NAB: "An Introduction to DTV Data Broadcasting," *NAB TV TechCheck*, National Association of Broadcasters, Washington, D.C., August 2, 1999.

8. "Navigation of DTV Data Broadcasting Services," *NAB TV TechCheck*, National Association of Broadcasters, Washington, D.C., November 1, 1999.

9. SMPTE Standard: SMPTE 325M-1999, "Opportunistic Data Broadcast Flow Control," SMPTE, White Plains, N.Y., 1999.

10. SMPTE Recommended Practice RP 203 (Proposed): "Real Time Opportunistic Data Flow Control in an MPEG-2 Transport Emission Multiplex," SMPTE, White Plains, N.Y., 1999.

11. ATSC: "ATSC Data Broadcast Standard," Advanced Television Systems Committee, Washington, D.C., Doc. A/90, July 26, 2000.

12. Chernock, Richard: "Implementation Recommendations for Data Broadcast," *NAB Broadcast Engineering Conference Proceedings*, National Association of Broadcasters, Washington, D.C., pp. 315–322, April 2000.

3.6 Bibliography

ATSC: "Amendment No. 1 to ATSC Standard: Program and System Information Protocol for Terrestrial Broadcast and Cable," Doc. A/67, ATSC, Washington, D.C, December 17, 1999.

ATSC: "Implementation of Data Broadcasting in a DTV Station," Advanced Television Systems Committee, Washington, D.C., Doc. IS/151, November 1999.

FCC Report and Order: "Closed Captioning Requirements for Digital Television Receivers," Federal Communications Commission, Washington, D.C., ET Docket 99-254 and MM Docket 95-176, adopted July 21, 2000.

SMPTE Standard: SMPTE 333M-1999, "DTV Closed-Caption Server to Encoder Interface," SMPTE, White Plains, N.Y., 1999.

The DVB System

Jerry C. Whitaker, Editor

4.1 Level One

The European Digital Video Broadcasting (DVB) program officially began in September 1993. Developmental work in digital television, then already under way in Europe, moved forward under this new umbrella. Meanwhile, a parallel activity, the *Working Group on Digital Television*, prepared a study of the prospects and possibilities for digital terrestrial television in Europe.

By 1999, a bit of a watershed year for digital television in general, the Digital Video Broadcasting Project had grown to a consortium of over 200 broadcasters, manufacturers, network operators, and regulatory bodies in more than 30 countries worldwide. Numerous broadcast services using DVB standards were operational in Europe and elsewhere around the world.

At the 1999 NAB Convention in Las Vegas, mobile and fixed demonstrations of the DVB system were made using a variety of equipment in various typical situations. Because mobile reception is the most challenging environment for television, the mobile system received a good deal of attention. DVB organizers used the demonstrations to point out the strengths of their chosen modulation method, the multicarrier *coded orthogonal frequency division multiplexing* (COFDM) technique.

In trials held in Germany beginning in 1997, DVB-T, the terrestrial transmission mode, had been tested in slow-moving city trams and high-speed inter-city trains. Following these successful trials, roll-out of the DVB-T system began.

4.2 Digital Video Broadcasting (DVB)

In the early 1990s, it was becoming clear that the once state-of-the-art MAC systems would have to give way to all-digital technology. DVB provided a forum for gathering all the major European television interests into one group to address the issue [3]. The DVB project promised to develop a complete digital television system based on a unified approach.

As the DVB effort was taking shape, it was clear that digital satellite and cable television would provide the first broadcast digital services. Fewer technical problems and a simpler regulatory climate meant that these new technologies could develop more rapidly than terrestrial systems. Market priorities dictated that digital satellite and cable broadcasting systems would have to be developed rapidly. Terrestrial broadcasting would follow later.

From the beginning, the DVB effort was aimed primarily at the delivery of digital video signals to consumers. Unlike the 1125-line HDTV system and the European HDTV efforts that preceded DVB, the system was not envisioned primarily as a production tool. Still, the role that the DVB effort played—and continues to play—in the production arena is undeniable. The various DVB implementations will be examined in this chapter.

4.2.1 Technical Background of the DVB System

From the outset, it was clear that the sound- and picture-coding systems of ISO/IEC MPEG-2 should form the audio- and image-coding foundations of the DVB system [3]. DVB would need to add to the MPEG transport stream the necessary elements to bring digital television to the home through cable, satellite, and terrestrial broadcast systems. Interactive television also was examined to see how DVB might fit into such a framework for new video services of the future.

MPEG-2

The video-coding system for DVB is the international MPEG-2 standard. As discussed previously in this book, MPEG-2 specifies a data-stream *syntax*, and the system designer is given a "toolbox" from which to make up systems incorporating greater or lesser degrees of sophistication [3]. In this way, services avoid being overengineered, yet are able to respond fully to market requirements and are capable of evolution.

The sound-coding system specified for all DVB applications is the MPEG audio standard MPEG Layer II (MUSICAM), which is an audio coding system used for many audio products and services throughout the world. MPEG Layer II takes advantage of the fact that a given sound element will have a masking effect on lower-level sounds (or on noise) at nearby frequencies. This is used to facilitate the coding of audio at low data rates. Sound elements that are present, but would not be heard even if reproduced faithfully, are not coded. The MPEG Layer II system can achieve a sound quality that is, subjectively, very close to the compact disc. The system can be used for mono, stereo, or multilingual sound, and (in later versions) surround sound.

The first users of DVB digital satellite and cable services planned to broadcast signals up to and including MPEG-2 Main Profile at Main Level, thus forming the basis for first-generation European DVB receivers. Service providers, thus, were able to offer programs giving up to "625-line studio quality" (ITU-R Rec. 601), with either a 4:3, 16:9, or 20:9 aspect ratio.

Having chosen a given MPEG-2 *compliance point*, the service provider also must decide on the operating bit rates (variable or constant) that will be used. In general, the higher the bit rate, the greater the proportion of transmitted pictures that are free of coding artifacts. Nevertheless, the law of diminishing returns applies, so the relationship of bit rate to picture quality merits careful consideration.

To complicate the choice, MPEG-2 encoder design has a major impact on receiver picture quality. In effect, the MPEG-2 specification describes only syntax laws, thus leaving room for technical-quality improvements in the encoder. Early tests by DVB partners established the approximate relationship between bit rate and picture quality for the Main Profile/Main Level, on the basis of readily available encoding technology. These tests suggested the following:

- To comply with ITU-R Rec. 601 "studio quality" on all material, a bit rate of up to approximately 9 Mbits/s is required.

- To match current "NTSC/PAL/SECAM quality" on most television material, a bit rate of 2.5 to 6 Mbits/s is required, depending upon the program material.

- Film material (24/25 pictures/s) is easier to code than scenes shot with a video camera, and it also will look good at lower bit rates.

MPEG-2 Data Packets

MPEG-2 *data packets* are the basic building blocks of the DVB system [3]. The data packets are fixed-length containers with 188 bytes of data each.

MPEG includes *program-specific information* (PSI) so that the MPEG-2 decoder can capture and decode the packet structure. This data, transmitted with the pictures and sound, automatically configures the decoder and provides the synchronization information necessary for the decoder to produce a complete video signal. MPEG-2 also allows a separate *service information* (SI) system to complement the PSI.

4.2.2 DVB Services

DVB has incorporated an *open service information system* to accompany the DVB signals, which can be used by the decoder and the user to navigate through an array of services offered. The following sections detail the major offerings.

DVB-SI

As envisioned by the system planners of DVB, the viewer of the future will be capable of receiving a multitude (perhaps hundreds) of channels via the DVB *integrated receiver decoder* (IRD) [3]. These services could range from interactive television to near video-on-demand to specialized programming. To sort out the available offerings, the DVB-SI provides the elements necessary for the development of an *electronic program guide* (EPG).

Key data necessary for the DVB IRD to automatically configure itself is provided for in the MPEG-2 PSI. DVB-SI adds information that enables DVB IRDs to automatically tune to particular services and allows services to be grouped into categories with relevant schedule information. Other information provided includes:

- Program start time

- Name of the service provider

- Classification of the event (sports, news, entertainment, and so on)

DVB-SI is based on four tables, plus a series of optional tables. Each table contains descriptors outlining the characteristics of the services/event being described. The tables are:

- *Network information table* (NIT). The NIT groups together services belonging to a particular network provider. It contains all of the tuning information that might be used during the setup of an IRD. It also is used to signal a change in the tuning information.

- *Service description table* (SDT). The SDT lists the names and other parameters associated with each service in a particular MPEG multiplex.

- *Event information table* (EIT). The EIT is used to transmit information relating to all the events that occur or will occur in the MPEG multiplex. The table contains information about the current transport and optionally covers other transport streams that the IRD can receive.

- *Time and date table* (TDT). The TDT is used to update the IRD internal clock.

In addition, there are three optional SI tables:

- *Bouquet association table* (BAT). The BAT provides a means of grouping services that might be used as one way an IRD presents the available services to the viewer. A particular service can belong to one or more "bouquets."

- *Running status table* (RST). The sections of the RST are used to rapidly update the running status of one or more events. The running status sections are sent out only once—at the time the status of an event changes. The other SI tables normally are repetitively transmitted.

- *Stuffing table* (ST). The ST may be used to replace or invalidate either a subtable or a complete SI table.

With these tools, DVB-SI covers the range of practical scenarios, facilitating a seamless transition between satellite and cable networks, near video-on-demand, and other operational configurations.

DVB-S

DVB-S is a satellite-based delivery system designed to operate within a range of transponder bandwidths (26 to 72 MHz) accommodated by European satellites such as the Astra series, Eutelsat series, Hispasat, Telecom series, Tele-X, Thor, TDF-1 and 2, and DFS [3].

DVB-S is a single carrier system, with the *payload* (the most important data) at its core. Surrounding this core are a series of layers intended not only to make the signal less sensitive to errors, but also to arrange the payload in a form suitable for broadcasting. The video, audio, and other data are inserted into fixed-length MPEG transport-stream packets. This packetized data constitutes the payload. A number of processing stages follow:

- The data is formed into a regular structure by inverting synchronization bytes every eighth packet header.

- The contents are randomized.

- Reed-Solomon forward error correction (FEC) overhead is added to the packet data. This efficient system, which adds less than 12 percent overhead to the signal, is known as the *outer code*. All delivery systems have a common outer code.

- Convolutional interleaving is applied to the packet contents.

- Another error-correction system, which uses a *punctured convolutional code*, is added. This second error-correction system, the *inner code*, can be adjusted (in the amount of overhead) to suit the needs of the service provider.

- The signal modulates the satellite broadcast carrier using quadrature phase-shift keying (QPSK).

In essence, between the multiplexing and the physical transmission, the system is tailored to the specific channel properties. The system is arranged to adapt to the error characteristics of the channel. Burst errors are randomized, and two layers of forward error correction are added. The second level (inner code) can be adjusted to suit the operational circumstances (power, dish size, bit rate available, and other parameters).

DVB-C

The cable network system, known as DVB-C, has the same core properties as the satellite system, but the modulation is based on quadrature amplitude modulation (QAM) rather than QPSK, and no inner-code forward error correction is used [3]. The system is centered on 64-QAM, but lower-level systems, such as 16-QAM and 32-QAM, also can be used. In each case, the data capacity of the system is traded against robustness of the data.

Higher level systems, such as 128-QAM and 256-QAM, are also possible, but their use depends on the capacity of the cable network to cope with the reduced decoding margin. In terms of capacity, an 8 MHz channel can accommodate a payload capacity of 38.5 Mbits/s if 64-QAM is used, without spillover into adjacent channels.

DVB-MC

The DVB-MC digital multipoint distribution system uses microwave frequencies below approximately 10 GHz for direct distribution to viewers' homes [3]. Because DVB-MC is based on the DVB-C cable delivery system, it will enable a common receiver to be used for both cable transmissions and this type of microwave transmission.

DVB-MS

The DVB-MS digital multipoint distribution system uses microwave frequencies above approximately 10 GHz for direct distribution to viewers' homes [3]. Because this system is based on the DVB-S satellite delivery system, DVB-MS signals can be received by DVB-S satellite receivers. The receiver must be equipped with a small microwave multipoint distribution system (MMDS) frequency converter, rather than a satellite dish.

DVB-T

DVB-T is the system specification for the terrestrial broadcasting of digital television signals [3]. DVB-T was approved by the DVB Steering Board in December 1995. This work was based on a set of user requirements produced by the Terrestrial Commercial Module of the DVB project. DVB members contributed to the technical development of DVB-T through the DTTV-SA (Digital Terrestrial Television—Systems Aspects) of the Technical Module. The European Projects SPECTRE, STERNE, HD-DIVINE, HDTVT, dTTb, and several other organizations developed system hardware and produced test results that were fed back to DTTV-SA.

As with the other DVB standards, MPEG-2 audio and video coding forms the payload of DVB-T. Other elements of the specification include:

- A transmission scheme based on *orthogonal frequency-division multiplexing* (OFDM), which allows for the use of either 1705 carriers (usually known as *2k*), or 6817 carriers (*8k*). Concatenated error correction is used. The 2k mode is suitable for single-transmitter operation and for relatively small single-frequency networks with limited transmitter power. The 8k mode can be used both for single-transmitter operation and for large-area single-frequency networks. The guard interval is selectable.

- Reed-Solomon outer coding and outer convolutional interleaving are used, as with the other DVB standards.

- The inner coding (punctured convolutional code) is the same as that used for DVB-S.

- The data carriers in the *coded orthogonal frequency-division multiplexing* (COFDM) frame can use QPSK and different levels of QAM modulation and code rates to trade bits for ruggedness.

- Two-level hierarchical channel coding and modulation can be used, but hierarchical source coding is not used. The latter was deemed unnecessary by the DVB group because its benefits did not justify the extra receiver complexity that was involved.

- The modulation system combines OFDM with QPSK/QAM. OFDM uses a large number of carriers that spread the information content of the signal. Used successfully in DAB (digital audio broadcasting), OFDM's major advantage is its resistance to multipath.

Improved multipath immunity is obtained through the use of a *guard interval*, which is a portion of the digital signal given away for echo resistance. This guard interval reduces the transmission capacity of OFDM systems. However, the greater the number of OFDM carriers provided, for a given maximum echo time delay, the less transmission capacity is lost. But, certainly, a tradeoff is involved. Simply increasing the number of carriers has a significant, detrimental impact on receiver complexity and on phase-noise sensitivity.

Because of the multipath immunity of OFDM, it may be possible to operate an overlapping network of transmitting stations with a single frequency. In the areas of overlap, the weaker of the two received signals is similar to an echo signal. However, if the two transmitters are far apart, causing a large time delay between the two signals, the system will require a large guard interval.

The potential exists for three different operating environments for digital terrestrial television in Europe:

- Broadcasting on a currently unused channel, such as an adjacent channel, or broadcasting on a clear channel

- Broadcasting in a small-area *single-frequency network* (SFN)

- Broadcasting in a large-area SFN

One of the main challenges for the DVB-T developers is that the different operating environments lead to somewhat different optimum OFDM systems. The common 2k/8k specification has been developed to offer solutions for all (or nearly all) operating environments.

4.2.3 The DVB Conditional-Access Package

The area of *conditional access* has received particular attention within DVB [3]. Discussions were difficult and lengthy, but a consensus yielded a package of practical solutions. The seven points of the DVB conditional-access package are:

- Two routes to develop the market for digital television reception should be permitted to coexist: receivers incorporating a single conditional-access system (the *Simulcrypt* route), and receivers with a common interface, allowing for the use of multiple conditional-access systems (the *Multicrypt* route). The choice of route would be optional.

- The definition of a common scrambling algorithm and its inclusion, in Europe, in all receivers able to descramble digital signals. This provision enables the concept of the single receiver in the home of the consumer.

- The drafting of a Code of Conduct for access to digital decoders, applying to all conditional-access providers.

- The development of a common interface specification.

- The drafting of antipiracy recommendations.

- Agreement that the licensing of conditional-access systems to manufacturers should be on fair and reasonable terms and should not prevent the inclusion of the common interface.

- The conditional-access systems used in Europe should allow for simple *transcontrol*; for example, at cable headends, the cable operators should have the ability to replace the conditional-access data with their own data.

4.2.4 Multimedia Home Platform

In 1997, the DVB Project expanded its scope to encompass the *multimedia home platform*, MHP [4]. From a service and application point of view, enhanced broadcasting, interactive services, and Internet access were deemed to be important to the future of the DVB system. The intention was to develop standards and/or guidelines that would establish the basis for an unfragmented horizontal market in Europe with full competition in the various layers of the business chain. A crucial role was expected to be played by the *application programming interface* (API). A comprehensive set of user- and market-based commercial requirements were subsequently approved.

Since the DVB project was established in 1993, it has produced a large family of specifications for almost every aspect of digital broadcasting. These specifications were subsequently adopted through the European Tele-communications Standards Institute (ETSI) as formal European standards. In its first phase, the DVB project focused its standardization work on the broadcasting infrastructure. As documented previously in this chapter, a comprehensive set of standards was delivered, including broadcast transmission standards for different transport media, service information standards related to services and associated transport networks, and transport-related standards for interactive services using different types of return channels (cable, PSTN, ISDN, and so on). In early implementations, however, problems developed wherein different applications and set-top boxes used different APIs that were incompatible. An end-user wanting to have access to all the DVB services available would, thus, need to buy several set top boxes. This formed a considerable road block in building full confidence of consumers in the future of digital TV services.

The expansion of the DVB project focus to include the standardization of a multimedia home platform was a logical next step. Aimed squarely at achieving full convergence of consumer information and entertainment devices, the MHP comprises the home terminal (set top box, integrated television set, multimedia personal computer), its peripherals, and an in-home digital network. From an application point of view, such standardization should lead to advanced broadcasting with multimedia data applications arriving alongside conventional linear broadcasting, plus interactive services and Internet access capabilities.

For standard 6, 7 or 8 MHz TV channels, the DVB standard offers a data throughput potential of between 6 Mbits/s and 38 Mbits/s, depending on whether only a part of the channel or the full channel or transponder is used. DVB systems provide a means of delivering MPEG-2 transport streams via a variety of transmission media. These transport streams traditionally contain MPEG-2-compressed video and audio. The use by DVB of MPEG-based "data containers" opens the way for anything that can be digitized to occupy these containers.

Implementation Considerations

The DVB data broadcasting standard allow a wide variety of different, fully interoperable data services to be implemented [5]. Data-casting or Internet services would typically use a broadcasters' extra satellite transponder space to broadcast content into the home via the consumer's receiving dish. The desired content would then be directed to the consumer's PC via a coaxial cable interfaced with a DVB-compliant plug-in PC card. After decoding, it could be viewed on a browser, or saved on the PC's hard disk for later use.

Where there is a need to have two-way communications, the user could connect via the public network to a specific host computer, or to a specific Web site. At the subscriber end, conditional access components built into the PC card would integrate with the subscriber management system, allowing the broadcaster to track and charge for the data that each subscriber receives.

The wide area coverage offered by a single satellite footprint ensures that millions of subscribers could receive data in seconds from just one transmission. Because much of the infrastructure is already in place, little additional investment would be needed from both broadcasters and subscribers to take advantage of data broadcasts over satellite. With possible data rates of more than 30 Mbits/s per transponder, a typical CD-ROM could be transmitted to an entire continent in just under three minutes.

4.2.5 DVB Data Broadcast Standards

DVB defines a set of methods for encapsulating data inside MPEG-2 transport stream packets [6]. There are various methods defined, each designed to provide a flexible and efficient means for supporting a specific set of applications. For example, the *multiprotocol encapsulation* method is intended for interconnecting two networks operating under various protocols by providing the facility for addressing multiple receivers, as well as efficient segmentation and de-segmentation of packets with arbitrary sizes.

DVB defines the following basic protocols for data broadcasting:

- *Data piping*—provides a mechanism for inserting data directly into the payload of transport stream packets. The mechanisms for the fragmentation of data into packets, the reassembly, and data interpretation are privately defined by users.

- *Asynchronous datagrams*—a data streaming method to encapsulate data inside PES (packetized elementary stream) packets in which the data has neither intra-stream nor inter-stream timing requirements.

- *Synchronous streaming data*—a data streaming method to encapsulate data inside PES packets in which the data streams are characterized by a periodic interval between consecutive packets so that both maximum and minimum arrival times between packets are bounded. Synchronous streams have no strong timing association with other data streams.

- *Synchronized streaming data*—a data streaming method to encapsulate data inside PES packets in which the data stream has the same intra-stream timing requirements as the synchronous streaming protocol. In addition, synchronized streaming implies a strong timing association with other PES streams, such as video and audio streams.

- *Multi-protocol encapsulation*—a format that provides a mechanism for transporting packets defined by arbitrary protocols, such as IP, inside MPEG-2 transport stream packets. The addressing scheme covers *unicast*, *multicast*, and broadcast applications.

- *Data carousel*—a format for encapsulating data into MPEG-2 streams that allows the server to present a set of distinct data modules to a receiver by cyclically repeating the contents of the carousel. If the receiver wants to access a particular module from the data carousel, it simply waits for the next time that the data for the requested module is broadcast.

- *Object carousel*—a format for encapsulating data into MPEG-2 streams that provides the facility to transmit a structured group of objects from a broadcast server to a broadcast client using *directory objects*, *file objects*, and *stream objects*.

DAVIC Standards

The Digital Audio Visual Council (DAVIC) is a nonprofit organization created to develop standards for the delivery of interactive data services to cable modems and set-top boxes [6]. DAVIC specifications define the minimum tools and dynamic behavior required by digital audio-visual systems for end-to-end interoperability across countries, applications, and services. To achieve this interoperability, DAVIC specifications define the technologies and information flows to be used within and between major components of generic digital audio-visual systems.

DAVIC specification encompasses the entire architectural components needed for the delivery of interactive audio-visual services. These components are:

- The server

- Delivery system

- Service consumer systems (i.e., set top boxes)

The specification covers all information layers, from the physical layer, middle- and high-layer protocols, to the managed object classes. DAVIC specifies a *reference decoder model* that defines specific memory and behavior requirements of a set-top device without specifying the internal design of the unit. Other parameters specified by the standard include the following:

- A *virtual machine* for application execution.

- Standard for presentation.

- Set of API's that are accessible by applications to be executed at the set-top device.

- Interfaces, protocols, and tools for implementing security, billing, system control, system validation, and conformance/interoperability testing.

The DVB data broadcast standard complements the DAVIC documents by specifying MPEG-2 based transport and physical layers of the DAVIC standard. Taken together, the DVB data broadcast and DAVIC specifications provide a complete, end-to-end specifications for implementing data broadcasting and interactive services.

4.2.6 DVB and the ATSC DTV System

As part of the ATSC DTV specification package, a recommended practice was developed for use of the ATSC standard to ensure interoperability internationally at the transport level with the European DVB project (as standardized by the European Telecommunications Standards Institute, ETSI). Guidelines for use of the DTV standard are outlined to prevent conflicts with DVB transport in the areas of *packet identifier* (PID) usage and assignment of user private values for descriptor tags and table identifiers.

Adherence to these recommendations makes possible the simultaneous carriage of system/service information (SI) conforming to both the ATSC standard (ATSC A/65) and the ETSI ETS-300-486 standard. Such dual carriage of SI may be necessary when transport streams conforming to the ATSC standard are made available to receivers supporting only the DVB service information standard, or when transport streams conforming to the

DVB standard are made available to receivers supporting only the ATSC SI standard.

4.3 References

1. "HD-MAC Bandwidth Reduction Coding Principles," Draft Report AZ-11, International Radio Consultative Committee (CCIR), Geneva, Switzerland, January 1989.

2. "Conclusions of the Extraordinary Meeting of Study Group 11 on High-Definition Television," Doc. 11/410-E, International Radio Consultative Committee (CCIR), Geneva, Switzerland, June 1989.

3. Based on technical reports and background information provided by the DVB Consortium.

4. Luetteke, Georg: "The DVB Multimedia Home Platform," DVB Project technical publication, 1998.

5. Jacklin, Martin: "The Multimedia Home Platform: On the Critical Path to Convergence," DVB Project technical publication, 1998.

6. Sariowan, H.: "Comparative Studies Of Data Broadcasting," *International Broadcasting Convention Proceedings*, IBC, London, England, pp. 115–119, 1999.

5

Standardization Issues

Jerry Whitaker, Editor

5.1   Introduction

Broadly stated, industrial development is dependent to a large degree on the adoption of system and component standards to permit the exchange of products and services. Thus, operational interchangeability is a prime consideration in the formulation of standards. In a narrower view, a group of standards may be dedicated to a specific system to ensure that the system is designed, tested, and operated to meet: first, the user's requirements, second, the interface requirements of other interconnected systems, and third, the performance requirements of the overall system.

In order to permit the end user, both professional and consumer, to acquire products that meet the performance requirements dictated by the intended application, test procedures and performance standards are necessary. These may be by government edict in the case of operations such as broadcasting, or products for which either quality or safety for consumer protection may be a criterion.

In addition to government regulation, there exists a large body of technical information consisting of recommended standards and practices that represent the consensus among those involved in a particular industry. These recommendations and industry-sponsored standards represent "good engineering practice" that ensures practical manufacturability within the current state of the art and, by consent of those skilled in the art, may serve as the basis for the generation of regulatory action by appropriate government agencies.

5.2 Classification of Standards

Television radio standards can be classified into three broad categories:

- Spectrum allocation and radiation standards, which relate primarily to the use of the available radio frequency spectrum
- Signal generation and transmission standards, which describe the specific signal parameters used to convey pictorial and aural information
- Equipment performance standards, which relate to the apparatus used to generate the visual and aural signal components

These broad categories, in turn, can be divided into several operational and reference classifications:

- Standards setting limits of performance
- Configuration standards
- Methods of measurement
- Procedures for operating equipment
- Terminology

The most commonly recognized category is the configuration standard, which provides dimensional and operating limits for a device or a process. Such standards have provided interchangeable digital audio/video recordings, for example. Another category is the method for measurement, which is sometimes combined with a configuration standard and provides repeatable and reproduceable tests for a device or a process. Other categories include standard definitions, procedures, and performance levels. Such standards usually arise out of a need in the industry and are put into writing by several organizations (not always in concert with one another).

The following references provide additional information on digital television in general, and the ATSC standard in particular.

5.2.1 Video

ISO/IEC IS 13818-1, International Standard (1994), MPEG-2 Systems

ISO/IEC IS 13818-2, International Standard (1994), MPEG-2 Video

ITU-R BT.601-4 (1994), Encoding Parameters of Digital Television for Studios

SMPTE 274M-1995, Standard for Television, 1920 × 1080 Scanning and Interface

SMPTE 293M-1996, Standard for Television, 720 × 483 Active Line at 59.94 Hz Progressive Scan Production, Digital Representation

SMPTE 294M-1997, Standard for Television, 720 × 483 Active Line at 59.94 Hz Progressive Scan Production, Bit-Serial Interfaces

SMPTE 295M-1997, Standard for Television, 1920 × 1080 50 Hz, Scanning and Interface

SMPTE 296M-1997, Standard for Television, 1280 × 720 Scanning, Analog and Digital Representation, and Analog Interface

5.2.2 Audio

ATSC Standard A/52 (1995), Digital Audio Compression (AC-3)

AES 3-1992 (ANSI S4.40-1992), AES Recommended Practice for digital audio engineering—Serial transmission format for two-channel linearly represented digital audio data

ANSI S1.4-1983, Specification for Sound Level Meters

IEC 651 (1979), Sound Level Meters

IEC 804 (1985), Amendment 1 (1989), Integrating/Averaging Sound Level Meters

5.3 ATSC Digital Television Standards

The Advanced Television Systems Committee (ATSC) is an international, non-profit organization developing voluntary standards for digital television. The ATSC has over 200 member organizations representing the broadcast, broadcast equipment, motion picture, consumer electronics, computer, cable, satellite, and semiconductor industries.

5.3.1 ATSC Standards Documents

The following is a partial list of ATSC Standards and technical activities.

ATSC Digital Television Standard, *Document A/53*

The Digital Television Standard describes the system characteristics of the advanced television (ATV) system. The document and its normative annexes provide detailed specification of the parameters of the system including the video encoder input scanning formats and the pre-processing and compression parameters of the video encoder, the audio encoder input signal format and the pre-processing and compression parameters of the audio encoder, the service multiplex and transport layer characteristics and normative specifications, and the VSB RF/transmission subsystem. The system is modular in concept and the specifications for each of the modules are provided in the appropriate annex. This document includes Amendment No. 1 to Doc. A/53.

Guide to the use of the ATSC Digital Television Standard, *Document A/54*

This guide provides an overview and tutorial of the system characteristics of the advanced television (ATV) system defined by ATSC Standard A/53, *ATSC Digital Television.*

Digital Audio Compression (AC-3), *Document A/52*

This document specifies coded representation of audio information and the decoding process, as well as information on the encoding process. The coded representation specified is suitable for use in digital audio transmission and storage applications, and may convey from 1 to 5 full bandwidth audio channels, along with a low frequency enhancement channel. A wide range of encoded bit-rates is supported by this specification. Typical applications of digital audio compression are in satellite or terrestrial audio broadcasting, delivery of audio over metallic or optical cables, or storage of audio on magnetic, optical, semiconductor, or other storage media.

Standard for Coding 25/50 Hz Video, *Document A/63*

This document describes the characteristics for the video subsystem of a digital television system operating at 25 Hz and 50 Hz frame rates.

Transmission Measurement and Compliance Standard for DTV, *Document A/64 Rev. A*

This document describes methods for testing, monitoring, and measurement of the transmission subsystem intended for use in the digital television (DTV) system, including specifications for maximum out-of-band emissions, parameters affecting the quality of the inband signal, symbol error tolerance, phase noise and jitter, power, power measure, frequency offset, and stability. In addition, it describes the condition of the RF symbol stream upon loss of MPEG packets. (The ATSC approved a revision to this document on May 30, 2000, that includes the revised FCC DTV emission mask.

PSIP for Terrestrial Broadcast and Cable, *Document A/65 Rev A with Amendment No. 1*

The Program and System Information Protocol Standard provides a methodology for transporting digital television system information and electronic program guide data. The standard includes an amendment that provides new functionality known as *Directed Channel Change* (DCC). This new feature will allow broadcasters to tailor programming or advertising based upon parameters defined by the viewer such as: postal, zip or location code, program identifier, demographic category, and content subject category. Potential applications include customized programming services, commercials based upon demographics, and localized weather and traffic reports.

Conditional Access For Terrestrial Broadcast, *Document A/70*

This document defines a standard for the Conditional Access system for ATSC terrestrial broadcasting to enable broadcasters to fully utilize the capabilities of digital broadcasting. This standard is based, whenever possible, on existing open standards and defines the building blocks necessary to ensure interoperability. The ATSC CA module is replaceable; to ensure that ATSC hosts are protected against obsolescence as security is upgraded. This standard applies to all CA vendors that supply CA service on behalf of an ATSC service provider. An overview of the CA standard is given in Annex C. (This document includes an Amendment that the ATSC approved on May 30, 2000.)

Modulation and Coding Requirements for DTV Applications Over Satellite, *Document A/80*

This document defines a standard for modulation and coding of data delivered over satellite for digital television contribution and distribution applications. The data can be a collection of program material including video, audio, data, multimedia, or other material. It includes the ability to handle multiplexed bit streams in accordance with the MPEG-2 systems layer, but it is not limited to this format and makes provision for arbitrary types of data as well. QPSK, 8PSK and 16 QAM modulation modes are included, as well as a range of forward error correction techniques.

Data Broadcast, *Document A/90*

The ATSC Data Broadcast Standard defines protocols for data transmission compatible with digital multiplex bit streams constructed in accordance with ISO/IEC 13818-1 (MPEG-2 systems). The standard supports data services that are both TV program related and non-program related. Applications may include enhanced television, webcasting, and streaming video services. Data broadcasting receivers may include PCs, televisions, set-top boxes, or other devices. The standard provides mechanisms for download of data, delivery of datagrams, and streaming data.

DTV Application Software Environment (DASE)

The DASE standard will define a software layer (middleware) that allows programming content and applications to run on a common receiver platform. Interactive and enhanced applications need access to common receiver features in a platform-independent manner. The standard will provide enhanced and interactive content creators the specifications necessary to ensure that their applications and data will run uniformly on all brands and models of receivers. Manufacturers will be able to choose hardware platforms and operating systems for receivers, but provide the commonality necessary to support applications made by many content creators.

Interactive Services

This standard defines session level protocols carried over interaction channels associated with interactive services. The interaction channel may be one or two way and connects a user (operating through a DTV receiver) with some service provider. The ATSC Interactive Services Protocols are intended to operate on a variety of physical networks, by focusing on higher layer protocols but not addressing specific applications

VSB Enhancements

Responding to the evolving needs of broadcasters, ATSC has initiated a standards activity aimed at enhancing the VSB modulation specifications that are a part of the ATSC Digital Television Standard (A/53). The new standards proceeding is intended to give broadcasters additional flexibility, including the ability to transmit programming and data to portable and mobile receivers. Just getting underway as this book went to press, it will move forward in parallel with the DTV implementation process already well underway.

Implementation Subcommittee

The ATSC Implementation Subcommittee (IS) evaluates technical requirements, operational impacts, preferred operating methods, time frames, and cost impacts of various DTV implementation issues. Based upon this analysis, the Subcommittee identifies potential requirements for standards, recommended practices, or guidelines.

5.3.2 ATSC DTV Standards and Reference Documents

The following documents form the basis for the ATSC digital television standard.

Service Multiplex and Transport Systems

ATSC Standard A/52 (1995), Digital Audio Compression (AC-3)

ISO/IEC IS 13818-1, International Standard (1994), MPEG-2 Systems

ISO/IEC IS 13818-2, International Standard (1994), MPEG-2 Video

ISO/IEC CD 13818-4, MPEG Committee Draft (1994), MPEG-2 Compliance

System Information Standard

ATSC Standard A/52 (1995), Digital Audio Compression (AC-3)

ATSC Standard A/53 (1995), ATSC Digital Television Standard

ATSC Standard A/80 (1999), Modulation And Coding Requirements For Digital TV (DTV) Applications Over Satellite

ISO 639, Code for the Representation of Names of Languages, 1988

ISO CD 639.2, Code for the Representation of Names of Languages: alpha-3 code, Committee Draft, dated December 1994

ISO/IEC 10646-1:1993, Information technology—Universal Multiple-Octet Coded Character Set (UCS) — Part 1: Architecture and Basic Multilingual Plane

ISO/IEC 11172-1, Information Technology—Coding of moving pictures and associated audio for digital storage media at up to about 1.5 Mbit/s—Part 1: Systems

ISO/IEC 11172-2, Information Technology—Coding of moving pictures and associated audio for digital storage media at up to about 1.5 Mbit/s—Part 2: Video

ISO/IEC 11172-3, Information Technology—Coding of moving pictures and associated audio for digital storage media at up to about 1.5 Mbit/s—Part 3: Audio

ISO/IEC 13818-3:1994, Information Technology—Coding of moving pictures and associated audio—Part 3: Audio

ISO/CD 13522-2:1993, Information Technology—Coded representation of multimedia and hypermedia information objects—Part 1: Base notation

ISO/IEC 8859, Information Processing—8-bit Single-Octet Coded Character Sets, Parts 1 through 10

ITU-T Rec. H. 222.0 / ISO/IEC 13818-1:1994, Information Technology—Coding of moving pictures and associated audio—Part 1: Systems

ITU-T Rec. H. 262 / ISO/IEC 13818-2:1994, Information Technology—Coding of moving pictures and associated audio—Part 2: Video

ITU-T Rec. J.83:1995, Digital multi-programme systems for television, sound, and data services for cable distribution

ITU-R Rec. BO.1211:1995, Digital multi-programme emission systems for television, sound, and data services for satellites operating in the 11/12 GHz frequency range

Receiver Systems

47 CFR Part 15, FCC Rules

EIA IS-132, EIA Interim Standard for Channelization of Cable Television

EIA IS-23, EIA Interim Standard for RF Interface Specification for Television Receiving Devices and Cable Television Systems

EIA IS-105, EIA Interim Standard for a Decoder Interface Specification for Television Receiving Devices and Cable Television Decoders

Program Guide

ATSC Standard A/53 (1995), ATSC Digital Television Standard

ANSI/EIA-608-94 (1994), Recommended Practice for Line 21 Data Service

ISO/IEC IS 13818-1, International Standard (1994), MPEG-2 Systems

5.3.3 Program/Episode/Version Identification

ATSC Standard A/53 (1995), Digital Television Standard

ATSC Standard A/65 (1998), Program and System Information Protocol for Terrestrial Broadcast and Cable

ATSC Standard A/70 (1999), Conditional Access System for Terrestrial Broadcast

ATSC Standard A/90 (2000), ATSC Data Broadcast Standard

ISO/IEC IS 13818-1, International Standard (1994), MPEG-2 systems

5.4 DVB

The following documents form the basis of the DVB digital television standard.

5.4.1 General Documents

Digital Satellite Transmission Systems, ETS 300 421

Digital Cable Delivery Systems, ETS 300 429

Digital Terrestrial Broadcasting Systems, ETS 300 744

Digital Satellite Master Antenna Television (SMATV) Distribution Systems, ETS 300 473

Specification for the Transmission of Data in DVB Bitstreams, TS/EN 301 192

Digital Broadcasting Systems for Television, Sound and Data Services; Subtitling Systems, ETS 300 743

Digital Broadcasting Systems for Television, Sound and Data Services; Allocation of Service Information (SI) Codes for Digital Video Broadcasting (DVB) Systems, ETR 162

Multipoint Distribution Systems

Digital Multipoint Distribution Systems at and Above 10 GHz, ETS 300 748

Digital Multipoint Distribution Systems at or Below 10 GHz, ETS 300 749

Interactive Television

Return Channels in CATV Systems (DVB-RCC), ETS 300 800

Network-Independent Interactive Protocols (DVB-NIP), ETS 300 801

Interaction Channel for Satellite Master Antenna TV (SMATV), ETS 300 803

Return Channels in PSTN/ISDN Systems (DVB-RCT), ETS 300 802

Interfacing to PDH Networks, ETS 300 813

Interfacing to SDH Networks, ETS 300 814

Conditional Access

Common Interface for Conditional Access and Other Applications, EN50221

Technical Specification of SimulCrypt in DVB Systems, TS101 197

Interfaces

DVB Interfaces to PDH Networks, prETS 300 813

DVB Interfaces to SDH Networks, prETS 300 814

5.5   SMPTE Documents Relating to Digital Television

The following documents relating to digital television have been approved (or are pending at this writing) by the Society of Motion Picture and Television Engineers.

5.5.1 General Topics

AES/EBU Emphasis and Preferred Sampling Rate, EG 32

Alignment Color Bar Signal, EG 1

Audio: Linear PCM in MPEG-2 Transport Stream, SMPTE 302M

Camera Color Reference Signals, Derivation of, RP 176-1993

Color, Equations, Derivation of, RP 177

Color, Reference Pattern, SMPTE 303M

Wide-Screen Scanning Structure, SMPTE RP 199

5.5.2 Ancillary

AES/EBU Audio and Auxiliary Data, SMPTE 272M

Camera Positioning by Data Packets, SMPTE 315M

Data Packet and Space Formatting, SMPTE 291M

DTV Closed-Caption Server to Encoder Interface, SMPTE 333M

Error Detection and Status Flags, RP 165

Format for Non-PCM Audio and Data in an AES3 Serial Digital Audio Interface, SMPTE 337M

Format for Non-PCM Audio and Data in an AES3 Serial Digital Audio Interface—ATSC A/52 (AC-3) Data Type, SMPTE 340M

Format for Non-PCM Audio and Data in an AES3 Serial Digital Audio Interface—Captioning Data Type, SMPTE 341M

Format for Non-PCM Audio and Data in an AES3 Serial Digital Audio Interface—Data Types, SMPTE 338M

Format for Non-PCM Audio and Data in an AES3 Serial Digital Audio Interface—Generic Data Types, SMPTE 339M

HDTV 24-bit Digital Audio, SMPTE 299M

LTC and VITC Data as HANC Packets, RP 196

Time and Control Code, RP 188

Transmission of Signals Over Coaxial Cable, SMPTE 276M

5.5.3 Digital Control Interfaces

Common Messages, RP 172

Control Message Architecture, RP 138

Electrical and Mechanical Characteristics, SMPTE 207M

ESlan Implementation Standards, EG 30

ESlan Remote Control System, SMPTE 275M

ESlan Virtual Machine Numbers, RP 182

Glossary, Electronic Production, EG 28

Remote Control of TV Equipment, EG 29

Status Monitoring and Diagnostics, Fault Reporting, SMPTE 269M

Status Monitoring and Diagnostics, Processors, RP 183-1995

Status Monitoring and Diagnostics, Protocol, SMPTE 273M

Supervisory Protocol, RP 113

System Service Messages, RP 163

Tributary Interconnection, RP 139

Type-Specific Messages, ATR, RP 171

Type-Specific Messages, Routing Switcher, RP 191

Type-Specific Messages, VTR, RP 170

Universal Labels for Unique ID of Digital Data, SMPTE 298M

Video Images: Center, Aspect Ratio and Blanking, RP 187

Video Index: Information Coding, 525- and 625-Line, RP 186

5.5.4 Edit Decision Lists

Device Control Elements, SMPTE 258M

Storage, 3-1/2-in Disk, RP 162

Storage, 8-in Diskette, RP 132

Transfer, Film to Video, RP 197

5.5.5 Image Areas

8 mm Release Prints, TV Safe Areas, RP 56

16 mm and 35 mm Film and 2 × 2 slides, SMPTE 96

Review Rooms, SMPTE 148

Safe Areas, RP 27.3

5.5.6 Interfaces and Signals

12-Channel for Digital Audio and Auxiliary Data, SMPTE 324M

Checkfield, RP 178

Development of NTSC, EG 27

Key Signals, RP 157

NTSC Analog Component 4:2:2, SMPTE 253M

NTSC Analog Composite for Studios, SMPTE 170M

Pathological Conditions, EG 34

Reference Signals, 59.94 or 50 Hz, SMPTE 305M

Bit-Parallel Interfaces

1125/60 Analog Component, RP 160

1125/60 Analog Composite, SMPTE 240M

1125/60 High-Definition Digital Component, SMPTE 260M

NTSC Digital Component, SMPTE 125M

NTSC Digital Component, 16 × 9 Aspect Ratio, SMPTE 267M

NTSC Digital Component 4:4:4:4 Dual Link, RP 175

NTSC Digital Component 4:4:4:4 Single Link, RP 174

NTSC Digital Composite, SMPTE 244M

Bit-Serial Interfaces

4:2:2p and 4:2:0p Bit Serial, SMPTE 294M

540 Mbits/s Serial Digital Interface, SMPTE 344M

Digital Component 4:2:2 AMI, SMPTE 261M

Digital Component S-NRZ, SMPTE 259M

Digital Composite AMI, SMPTE 261M

Digital Composite, Error Detection Checkwords/Status Flag, RP 165

Digital Composite, Fiber Transmission System, SMPTE 297M

Digital Composite, S-NRZ, SMPTE 259M

Element and Metadata Definitions for the SDTI-CP, SMPTE 331M

Encapsulation of Data Packet Streams over SDTI (SDTI-PF), SMPTE 332M

HDTV, SMPTE 292M

High Data-Rate Serial Data Transport Interface (HD-SDTI), SMSPTE 348M

HDTV, Checkfield, RP 198

Jitter in Bit Serial Systems, RP 184

Jitter Specification, Characteristics and Measurements, EG 33

Jitter Specification, Measurement, RP 192

SDTI Content Package Format (SDTI-CP), SMPTE 326M

Serial Data Transport Interface, SMPTE 305M

Time Division Multiplexing Video Signals and Generic Data over High-Definition Interfaces, SMPTE 346M

Vertical Ancillary Data Mapping for Bit-Serial Interface, SMPTE 334M

Scanning Formats

1280 × 720 Scanning, SMPTE 296M

1920 × 1080 Scanning, 60 Hz, SMPTE 274M

1920 × 1080 Scanning, 50 Hz, SMPTE 295M

720 × 483 Digital Representation, SMPTE 293M

5.5.7 Monitors

Alignment, RP 167

Colorimetry, RP 145

Critical Viewing Conditions, RP 166

Receiver Monitor Setup Tapes, RP 96

5.5.8 MPEG-2

4:2:2 Profile at High Level, SMPTE 308M

4:2:2 P@HL Synchronous Serial Interface, SMPTE 310M

Alignment for Coding, RP 202

MPEG-2 Video Elementary Stream Editing Information, SMPTE 328M

MPEG-2 Video Recoding Data Set, SMPTE 327M

MPEG-2 Video Recoding Data Set—Compressed Stream Format, SMPTE 329M

Opportunistic Data Broadcast Flow Control, SMPTE 325M

Splice Points for the Transport Stream, SMPTE 312M

Transport of MPEG-2 Recoding Information as Ancillary Data Packets, SMPTE 353M

Transporting MPEG-2 Recoding Information Through 4:2:2 Component Digital Interfaces, SMPTE 319M

Transporting MPEG-2 Recoding Information Through High-Definition Digital Interfaces, SMPTE 351M

Unique Material Identifier (UMID), SMPTE 330M

5.5.9 Test Patterns

Alignment Color Bars, EG 1

Camera Registration, RP 27.2

Deflection Linearity, RP 38.1

Mid-Frequency Response, RP 27.5

Operational Alignment, RP 27.1

Safe Areas, RP 27.3

Telecine Jitter, Weave, Ghost, RP 27.4

5.5.10 Video Recording and Reproduction

Audio Monitor System Response, SMPTE 222M

Channel Assignments, AES/EBU Inputs, EG 26

Channel Assignments and Magnetic Masters to Stereo Video, RP 150

Cassette Bar Code Readers, EG 31-1995

Data Structure for DV-Based Audio, Data, and Compressed Video, SMPTE 314M

Loudspeaker Placement, HDEP, RP 173

Relative Polarity of Stereo Audio Signals, RP 148

Telecine Scanning Capabilities, EG 25

Tape Care, Handling, Storage, RP 103

Time and Control Code

Binary Groups, Date and Time Zone Transmissions, SMPTE 309M

Binary Groups, Storage and Transmission, SMPTE 262M

Directory Index, Auxiliary Time Address Data, RP 169

Directory Index, Dialect Specification of Page-Line, RP 179

Specifications, TV, Audio, Film, SMPTE 12M

Time Address Clock Precision, EG 35

Vertical Interval, 4:2:2 Digital Component, SMPTE 266M

Vertical Interval, Encoding Film Transfer Information, 4:2:2 Digital, RP 201

Vertical Interval, Location, RP 164

Vertical Interval, Longitudinal Relationship, RP 159

Vertical Interval, Switching Point, RP 168

5.6   SCTE Standards

The following documents relating to digital television have been adopted by the Society of Cable Telecommunications Engineers.

DVS 011 Cable and Satellite Extensions to ATSC System Information Standards

DVS 018 ATSC Digital Television Standard

DVS 019 Digital Audio Compression (AC-3) Standard

DVS 020 Guide to the Use of the ATSC Digital Television Standard

DVS 022 System Information for Digital Information

DVS 026 Subtitling Method for Broadcast Cable

DVS 031 Digital Video Transmission Standard for Cable Television

DVS 033 SCTE Video Compression Formats

DVS 043 QPSK Tools for Forward and Reverse Data Paths

DVS 046 Specifications for Digital Transmission Technologies

DVS 047 National Renewable Security Standard (NRSS)

DVS 051 Methods for Asynchronous Data Services Transport

DVS 053 VBI Extension for ATSC Digital Television Standards

DVS 055 EIA Interim Standard IS-679 B of National Renewable Security Standard (NRSS)

DVS 057 Usage of A/53 Picture (Video) User Data

DVS 061 SCTE Cable and Satellite Extensions to ATSC System Information (SI)

DVS 064 National Renewable Security Standards (NRSS) Part A and Part B

DVS 068 ITU-T Recommendation J.83–"Digital Multi-Programme Systems for Television, Sound and Data Services for Cable Distribution"

DVS 071 Digital Multi Programming Distribution by Satellite

DVS 076 Digital Cable Ready Receivers: Practical Considerations

DVS 077 Requirements for Splicing of MPEG-2 Transport Streams

DVS 080 Digital Broadband Delivery Phase 1.0 Functions

DVS 082 Broadband File System Product Description Release 1.2s

DVS 084 Common Interface for DVB Decoder Interface

DVS 085 DAVIC 1.2 Basic Security

DVS 092 Draft System Requirements for ATV Channel Navigation

DVS 093 Draft Digital Video Service Multiplex and Transport System

DVS 097 Program and System Information Protocol for Terrestrial Broadcast and Cable

DVS 098 IPSI Protocol for Terrestrial Broadcast with examples

DVS 110 Response to SCTE DVS CFI Cable Headend and Distribution Systems

DVS 111 Digital Headend and Distribution CFI Phase 1.0 System Description

DVS 114 SMPTE Splice point for MPEG-2 Transport

DVS 131 Draft Point-of-Development (POD) Proposal on Open Cable

DVS 132 Methods for Isochronous Data Services Transport

DVS 147 Revision to DVS 022 (Standard System Information)

DVS 151 Operational Impact on Currently Deployed Systems

DVS 153 ITU-T Draft Recommendation J.94

DVS 154 Digital Program Insertion Control API

DVS 157 SCTE Proposed Standard Methods for Carriage of Closed Captions and Non-Real Time Sampled Video

DVS 159 Optional Extensions for Carriage of NTSC VBI Data in Cable Digital Transport Streams

DVS 161 ATSC Data Broadcast Specification

DVS 165 DTV Interface Specification

DVS 166 Draft Corrigendum for Program ad System Information Protocol for Terrestrial Broadcast and Cable (A/65

DVS 167 Digital Broadband Delivery System: Out of Band Transport– Quadrature Phase Shifting Key (QPSK) Out of Band Channels Based on DAVIC, first draft

DVS 168 Emergency Alert System Interface to Digital Cable Network

DVS-178 Cable System Out-of-Band Specifications (GI)

DVS-179 MPAA Response to DVS CFI

DVS-181 Service Protocol

DVS-190 Standard for Conveying VBI Data in MPEG-2 Transport Streams

DVS-191 Draft Standard API to Splicing Equipment for MPEG-2 Transport StreamsDVS-192 Splicer Application Programmer's Interface Definition Overview

DVS-194 Home Digital Network Interface Specification Proposal with Copy Protection

DVS-195 Home Digital Network Interface Specification Proposal without Copy Protection

DVS-208 Proposed Standard: Emergency Alert Message for Cable

DVS-209 DPI System Physical Diagram

DVS-211 Service Information for Digital Television

DVS-213 Copy Protection for POD Module Interface

5.7 Standardization and Implementation Issues

Video standards have a complex structure [1]. They provide a detailed description of video signals compatible with an intended receiver. Strictly speaking, a broadcast television standard is a set of technical specifications defining the method of on-air radio-frequency transmission of a picture with accompanying sound. However, in the video program production environment, the RF parameters are usually irrelevant and sound can be handled in many different ways. For this reason, the term *video standard* is more appropriate.

It is possible, and many times useful, to divide a video standard into five component parts, as follows:

• The *scanning standard*, which determines how the picture is sampled in space and in time (i.e., the number of lines in the picture, the number of pictures per second, and whether interlace or progressive scanning is used). Only the conversion of video signals between different scanning standards—for example, NTSC to PAL—can truly be called *standards conversion*.

- *Color information representation*, which determines how color information is conveyed. Color formats are divided into two basic categories: *component*, primarily used for production, and *composite*, traditionally used for conventional television broadcasting.

- *Aspect ratio*, which describes how the picture fits into a screen of a particular proportion.

- *Signal levels*, which determine how a receiver will interpret a video waveform in the voltage domain. In the digital domain, it is necessary to standardize the relationship between analog voltages and the digital codes that they signify.

- *Format*, which is an agreed way of packaging the picture information for transmission or recording.

Historically, the most significant boundary in the television world was the one between different frame rates. At the dawn of black-and-white television, it was reasonable to link the field repetition rate with the frequency of the ac power line. This prevented slow scroll of the horizontal hum bar on the received picture. When the color era began, with crystal-referenced operating frequencies and improved filtering of the dc power supply, this link became more or less irrelevant. The only justifiable reason for preserving the relationship was to reduce the visibility of unpleasant low-frequency beating between studio lighting and the video camera field rate. Unfortunately, the economic (and political) necessity for backward compatibility did not permit significant changes to be made to the field and frame rates.

There are a number of organizations responsible for standardization at global, regional, and national levels. In the broadcast television field, one of the most prominent is the former CCIR—the International Radio Consultative Committee. CCIR was a branch of the International Telecommunication Union (ITU), which, in turn, is part of the United Nations Organization. In 1993, the CCIR was renamed as ITU-R (the Radiocommunication Sector of the ITU). Television is a main topic for ITU-R Study Group 11. The crucial documents issued by ITU-R are Recommendations and Reports. They contain explicit information, but in reasonably general form, often leaving room for different variants in practical implementation.

5.7.1 Digital Television System Model

In April 1997, the ITU adopted a series of Recommendations (standards) defining a *digital terrestrial television broadcasting* (DTTB) system [2]. The set of Recommendations and associated descriptive Reports represents a four-year effort by ITU Task Group 11/3. The Task Group began with the premise that it should establish an infrastructure that enabled a communication environment in which a continuum of television and other data services could be brought to the consumer via wire, recorded media, and through the air. Terrestrial broadcasting presents the most challenging set of constraints of all of the forms of media. Therefore, a system that works well in the terrestrial broadcasting environment should suffice for other media. The first step in harmonizing the delivery of these services was the development of a suitable service model.

The Task Group work was based in many ways on both the work of and the philosophy behind the MPEG-2 standard. The MPEG-2 document provides a set of tools that can be used to describe a system. The set of Recommendations developed by Task Group 11/3 defined a constrained set of tools that can be used to provide a DTTB service. This set of tools provides for a single, low-cost decoder that can deliver both ATSC- and DVB-coded images and sound.

Task Group 11/3 also established a harmonized subset of MPEG-2 that allowed for a single decoder that can translate the service multiplex and transport layer, and decode the audio and video compression and coding layers for any system that conforms to the DTTB set of Recommendations.

The set of harmonized Recommendations fully meets the request of the World Broadcasting Union for unique global broadcasting systems leading to single universal consumer appliances. A report on the Task Group's efforts can be found in [2].

5.7.2 Harmonized Standards

The convergence of the television, computer, and communications industries is well under way, having been anticipated for quite some time [3]. Video and audio compression methods, server technology, and digital networking are all making a significant impact on television production, post-production, and distribution. Accompanying these technological changes are potential benefits in reduced cost, improved operating efficiencies and creativity, and increased marketability of material. Countering the potential benefits are threats of confusion, complexity, variable technical performance, and increased costs if not properly managed. The technological changes now unfolding will dramatically alter the way in which video is produced and distributed in the future.

In this context, the Society of Motion Picture and Television Engineers (SMPTE) and the European Broadcasting Union (EBU) jointly formed the *Task Force for Harmonized Standards for the Exchange of Program Material as Bitstreams*. The Task Force, with the participation by approximately 200 experts from around the world, produced two reports. The first, published in April 1997, was called *User Requirements* for the systems and techniques that will implement new technologies. The second report, published in July 1998, provided *Analyses and Results* from the deliberations of the Task Force. Taken together, the two documents are meant to guide the converging industries in their decisions regarding specific implementations of the applicable technologies, and to steer the future development of standards that are intended to maximize the benefits and minimize the detriments of implementing such systems.

The goals of the Task Force were to look into the future a decade or more, to determine the requirements for systems in that time frame, and to identify the technologies that could be implemented within the coming years in order to meet the specified requirements over the time period. This approach recognized that it takes many years for new technologies to propagate throughout the industries implicated in the changes. Because of the large and complex infrastructures involved, choices must be made of the methods that can be applied in the relatively near future, but which will still be viable over the time period contemplated by the Task Force's efforts.

To meet its objectives, the Task Force partitioned its work among six separate Sub-Groups, each of which was responsible for a portion of the investigation. These Sub-Groups included the following general subject areas:

- Systems

- Compression

- Wrappers and File Formats

- Metadata

- File Transfer Protocols

- Physical Link and Transport Layers for Networks

Some of the Sub-Groups found that their areas of interest were inextricably linked with one another and, consequently, performed their work jointly (and produced a common report).

The Task Force made significant contributions in identifying the technologies and standards required to carry converging industries with an interest in television to the next plane of cooperation and interoperation. A detailed report of the Task Force can be found in [3].

5.7.3 Top Down System Analysis

In an effort to facilitate the transition to digital television broadcasting, the Implementation Subcommittee (IS) of the ATSC undertook an effort to inventory the various systems and their interfaces that could potentially exist in a typical station, regardless of the implementation scenario. This inventory was intended to serve as a guide to point to the standards that exist for equipment interfaces, identity potential conflicts between those standards, and identity areas where standards and/or technology need further development.

The resulting report and referenced system maps were generalized blueprints for not only construction of an early DTV facility, but also a joint informal agreement among a number of industry manufacturers and end-user consultants, all of whom have worked in digital television for some time. For the station engineer, the maps provide a basic blueprint for their facilities. No station would build a facility as shown in the main system map, but rather each would take portions of the map as a foundation upon which to build to meet their local requirements.

By establishing a single system map where all of the likely system elements and interfaces could exist, a commonality in design and functionality throughout the industry was established that not only pointed the way for the early DTV adopters, but also established a framework upon which to build and expand systems in the future.

One of the most interesting aspects of the committee's efforts was the single format philosophy of the *plant native format*.

Plant Native Format

Several DTV plant architectures were considered by the ATSC IS [4]. The most basic function, pass through of a pre-compressed stream, with or without local bit-stream splicing, is initially economical but lacks operational flexibility. To provide the operational flexibility required, digital video levels could be added to an existing analog NTSC facility, but managing multiple video formats would likely become complex and expensive. An alternative is to extend the existing facility utilizing a single-format plant core infrastructure surrounded by appropriate format conversion equipment. This single-format philosophy of facility design is based on application of one format, called the *plant native format*, to as much of the facility as practical.

The chosen native format would be based on numerous factors, including the preferred format of the station's network, economics, existing legacy equipment and migration strategy, equipment availability, and station group format.

A native format assumes that all material will be processed in the plant in a single common standard to allow production transitions such as keying, bug insertions, and so on. The format converters at the input of the plant would convert the contribution format to the native plant format. A format converter would not be necessary if the native format is the same as the contribution format. Legacy streams will also likely require format conversion. The native plant infrastructure could be traditional analog or digital routers, production switchers, master control switchers, or new concepts such as servers and digital networks.

Choosing a native plant format has a different set of criteria than choosing a broadcast emission format. Input signals arrive in a variety of formats, necessitating input signal format conversion. Similarly, changes in broadcast emission format will necessitate output format conversion. Therefore the native plant format should be chosen to help facilitate low-cost, low-latency, and high-quality format conversions. It is well understood that format conversions involving interlaced scan formats require careful attention to de-interlacing, while progressive scan formats offer fewer conversion challenges.

The native plant format could be chosen from a variety of candidates, including:

- 480I 4:3 or 16:9 carried by SMPTE 259M-1997

- 1080I or 720P using SMPTE 292M

- 480P using SMPTE 293M-1996

- Intermediate compressed formats

Although the full benefits of adopting a plant native format are most apparent with an implementation in which "plant native format" means a single plant native format, many stations will not be able to afford to adopt a single plant native format immediately. For them "plant native format" will, at least on an interim basis, mean a practical mix of perhaps two formats. One of the formats might be the legacy format, which is being phased out while the other is the preferred format for future operations.

5.7.4 Advanced System Control Architecture

The early experiences of DTV facility construction demonstrated the need for a comprehensive control and monitoring environment for studio equipment [5]. Because of differing product lines and market directions, various systems were derived from different sets of operational requirements. Furthermore, varied enabling technologies within the network and computing industries were incorporated into these systems. In recognition of this somewhat hap-hazard situation, and the need for broadcasters and audio/video professionals to chart the course, rather than just follow, a working group within the SMPTE was formed to study advanced system control architectures.

The focus of the S22.02 Advanced System Control Architectures Working Group was to unify existing systems and proposed new systems. This work was intended to be a living document, specifically:

- To provide a comprehensive system overview

- Identify specific areas for additional SMPTE efforts

- Solicit input from interested parties

At this writing, the SMPTE was preparing to embark on standardization efforts in this area.

In terms of overall architecture, the Working Group's efforts were intended to meet the following requirements:

- To be sufficiently flexible to accommodate varying enabling technologies where possible

- Independent of specific technologies where possible

- Scalable to a range of platforms and environments

- Extensible to adapt to emerging technologies

- Modular in nature, allowing functional pieces of the system to be used as building blocks

- Offer a viable migration path for existing facilities

Functional Planes

Under the S22.02 Working Group plan, *functional planes* exist as the front line interface between the studio applications and the functionality of the control system [5]. Four basic functional planes are defined:

- **Content management**: Manages content in the studio; understands physical storage allocation for the content; performs activities including content distribution, content creation, scheduled operations, and storage management; presents a view of content as required for *data streaming*

- **Service**: Combines multiple content streams into complete services; maps services onto resources available in the *path plane*; abstracts the mapping of content to individual pieces of equipment; uses content sources to create paths

- **Path**: Facilitates the physical connection between devices for the purposes of data streaming; manages the physical links between devices; manages the resources required for these connections

- **Device**: Contains the interfaces used to access studio equipment; provides specific information for device I/O capability (ports); is based on a SMPTE-defined hierarchy of functional classifications; provides for extensible interfaces

The planes are organized into a form of *usage abstraction* from content, to service, to path, to device. Adjacent planes cooperate to carry out individual operations as generated by the studio applications. Though this abstraction is beneficial in most cases, it is not explicitly required. For example, applications can access functionality at any plane, or planes can access any other plane in order to accomplished a given task.

These functional planes represent logical partitioning within the control system. From an implementation point of view, the devices themselves will likely be responsible for running local software that will contribute to the functionality of one or more of these planes. For example, within a network router, software exists at runtime that is responsible for some path plane functionality.

Content Management Plane

The *content plane* is the locus of all the "higher level" applications ("business" applications) in the studio and is the first line of support for these applications. [5] It implements *content management* classes of objects, including:

- **Library server**. A studio will have zero or more objects of this class, whose purpose is to provide a *library service*. If there are multiple such objects, they will link together to provide a unified, though distributed, library service.

- **Streaming element**. Objects of this class are created to present a view of content files as endpoints for streaming file transfers. These objects will be created as necessary to support scheduled operations, and be deleted after the operations end.

Service Plane

The *service plane* is responsible for aggregating multiple content streams into complete services and then mapping the implementation of these services onto aggregated resources exposed by the path plane [5]. The service plane manages multiple logical paths within the network that together represent a program, and therefore need to be treated together.

Path Plane

The *path plane* represents the end-to-end connectivity from one device to another, or through chained devices, over a network of routers [5]. The responsibility of the path plane is to transport the media content from one point to another to accomplish a particular task in a timely fashion and with a guaranteed quality of service.

A *physical connectivity* path is a connection between a source device and a destination device via a number of intermediate devices, i.e. a number of chained devices. The adjacent devices are connected through a network connection that is routed over the routers in the network. Control and management of the physical connectivity path involves allocation, setup, and monitoring of the components (devices and links between adjacent devices) that make up the path.

A logical and abstract representation of the connectivity path is desirable to simplify the interface for control of the content data flow. The concept of a *stream* is used to represent such a logical view and the controllable attributes of the content data transfer process. A stream is a logical construction (object) corresponding to the transient state of content data being passed from one device to another through the physical connectivity path.

The path plane contains a collection of logical entities (objects) that collectively build, operate, and manage the connectivity. These entities can be categorized into two functional groups:

- Streams that control the content data flow over the path

- *Studio resource management* (SRM) that facilitates and manages the allocation and reservation of shared studio resources to meet the resource requirements of various tasks and thereby achieve better resource utilization. Manageable studio resources include the devices that are registered with the device *registration repository* and the network connections and bandwidth. SRM can also provide a central point for resource *access control*. The SRM functions may be carried out collaboratively by a number of logical entities.

Device Plane

The *device plane* represents the collection of object classes that implement the interfaces used to access equipment throughout the studio [5]. Devices can also represent the interfaces to software-only objects that perform specific functions within the control network. These device objects need to support the following generic functionality:

- *Dynamic discovery* of the various *interface elements* used in the interface. This does not necessarily mean that the interface description information travels over any network; it may be derived from a local information database for the devices.

- Interfaces for later versions of a device are compatible with earlier versions of the interface.

- Interface elements are based on a consistent set of SMPTE-defined broadcast types.

- All information that can be used to define the unambiguous operational state of a device is available in the device interface. That is, a device state can be completely represented (and therefore restored) by a snapshot of its current interface state.

- Device interfaces are made up of a hierarchical collection of sub-functions (functional blocks).

- Interfaces are extensible and therefore inherently backward compatible.

Functional Layers

Functional layers exist as the foundation on which the planes are built [5]. They provide a homogeneous set of functions that facilitate the transport of control information throughout the network. The three basic functional layers are:

- **System**: Provides common reusable system services as follows: studio time, configuration, security, identification and naming, fault recovery, and status and alarming.

- **Communication**: Provides a common facility for the exchange of information among communicating endpoints within a distributed network, allows the use of multiple protocols simultaneously, and facilitates the transfer of time-synchronized operations.

- **Transport**: Provides for the reliable transmission of information throughout a network, provides an abstraction for physical network access, and allows for adaptation to legacy networks.

5.8 References

1. Pank, Bob (ed.): *The Digital Fact Book*, 9th ed., Quantel Ltd, Newbury, England, 1998.

2. Baron, Stanley: "International Standards for Digital Terrestrial Television Broadcast: How the ITU Achieved a Single-Decoder World," *Proceedings of the 1997 BEC*, National Association of Broadcasters, Washington, D.C., pp. 150–161, 1997.

3. SMPTE and EBU, "Task Force for Harmonized Standards for the Exchange of Program Material as Bitstreams," *SMPTE Journal*, SMPTE, White Plains, N.Y., pp. 605–815, July 1998.

4. ATSC: "Implementation Subcommittee Report on Findings," Draft Version 0.4, ATSC, Washington, D.C., September 21, 1998.

5. SMPTE: "System Overview—Advanced System Control Architecture, S22.02, Revision 2.0," S22.02 Advanced System Control Architectures Working Group, SMPTE, White Plains, N.Y., March 27, 2000.

5.9 Bibliography

Battison, John: "Making History," *Broadcast Engineering*, Intertec Publishing, Overland Park, Kan., June 1986.

Benson, K. B., and Jerry C. Whitaker (eds.): *Television Engineering Handbook*, rev. ed., McGraw-Hill, New York, N.Y., 1992.

Benson, K. B., and J. C. Whitaker (eds.): *Television and Audio Handbook for Engineers and Technicians*, McGraw-Hill, New York, N.Y., 1989.

McCroskey, Donald: "Setting Standards for the Future," *Broadcast Engineering*, Intertec Publishing, Overland Park, Kan., May 1989.

6

Metadata Management

Mark Grossman, Geocast Network Systems

6.1   Introduction

Metadata is a key element in making interactive television—or any form of data broadcasting—work. Put simply, metadata is digital information that describes other data. It can simply describe the type and size of a data file, or the start and stop times of a scene within an encoded program. At the other extreme, it can be a detailed description of the contents of a digitized video, such as which themes, actions, people and objects appear (and disappear). Pieces of metadata are often referred to as *tags* or *attributes*.

6.1.1 Metadata Defined

Metadata itself is a form of data that can be stored as part of a file header. It can also reside separately in a database, a search engine, or distributed around the Internet. What is important is that it can be economically broadcast along with the material it describes.

The structure and meaning of metadata can be described in a form that people and systems can use to process metadata tags. This description of metadata is called a *schema*. This term is often used in traditional database management to refer to the meaning of and relationships among tables and table columns. Metadata is important because it helps humans and machines understand the information being broadcast and presented.

In the authoring domain, creators need to identify, edit, and track elements when they make content. Stored with an edited sequence or alongside it, metadata allows content creators to find, build on, and recover source material. Content can be uniquely identified, and then organized by any desired attribute.

Broadcasters and media companies often have a production automation and archival challenge, their goal being to reduce the cost of moving media around, doing research, managing flow, and repurposing for Internet and other broadband distribution outlets. The benefit is more revenue per foot of tape or film. This problem is being addressed by providers of database-centric media management systems and services, and by sellers of new breeds of newsroom editing systems and broadcast servers.

These kinds of systems are not necessarily ideal for data broadcasting, but as more of them go on line and are integrated into a total production and broadcast network, the metadata they use will become more practical for data broadcasting. To some extent, the same kind of metadata that allows a journalist or producer to find and organize stories and clips into a one-hour broadcast can help a consumer find the stories he or she is searching for (although a typical slug like "exterior shot of mall entrance" may not be an ideal search key).

Another flavor of metadata aids in the mechanics of storing, transferring, and streaming media. It may describe items such as the encoding bit rate and broadcast times—attributes that help with the task of inserting material into the DTV stream. Still another type guides the presentation style, rights, and permissions for an end-user consuming the media. In the Web and data management worlds, *meta tags* and keywords drive world-wide search capabilities.

When the right set of features from all these domains is used, data broadcasting becomes practical. Metadata can be generally divided into two basic types:

- *Declarative* tags, which convey the general identification, administration, and form of an object—where it came from and where it is going. This includes basics such as a unique identifier, something to designate the type of media (audio, video, HTML, and so on), and the name of the content provider.

- *Interpretive* tags, which relate to what the object is about. This may include the genre or style of entertainment, or descriptive keywords. Given the right set of interpretive tags on material and a related set of tags on advertisements, the selection of a targeted ad can be automated.

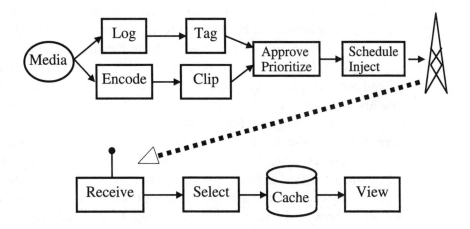

Figure 6.1 Typical broadcast data flow.

6.2   A Metadata-Driven System

Among the numerous benefits of datacasting to broadcasters and other information delivery operators are the ability to leverage existing DTV bandwidth to provide additional services and the ability to extend the value of existing content. A great value proposition for consumers is to marry the one-to-many delivery model and reliable high bandwidth of television with the control and customized access model of the Internet. This approach uses a portion of the digital broadcast spectrum available at DTV-capable TV stations to deliver 30 f/s, full-resolution video, CD-quality audio, and data feeds to the desktop. The user plugs in a receiver as an external peripheral to the PC. The receiver is an always-on device with a large disk cache; the user can then access a personalized mix of television highlights streamed from the receiver, as well as digital music, short-form entertainment, and software downloads. The system is integrated with a standard Internet connection for applications such as e-commerce fulfillment, user feedback, and other material. The emphasis is on near-real-time original and repurposed content—media and information presented in a readable, *lean-forward* environment.

Figure 6.1 shows a detailed view of how digital content and metadata flow through the broadcast system. Media can be near-live or on-air video; it can be taped ads, MP3 audio files, CDs, or packaged digital material such as on-line catalogs or text feeds. The bottom fork is taken by the *essence*, the actual bits to be presented to the user. Digitized in an appropriate format for broadcast, this data stream is then marked into segments such as individual news clips or music tracks. The top fork in the diagram is the path taken by metadata, some of which is extracted from the media (i.e., logged) and the rest of which is entered in a tagging process. The resulting essence and associated metadata are then subject to an editorial or programming process—mainly a quality control process—where it can be categorized and prioritized and, thus, prepared for broadcast. The combined essence and metadata stream is demodulated by the end user's receiver. The metadata is submitted to a preference matching and valuation process, and if a piece of material meets the specified criteria, it is kept in the receiver's disk cache. The arrival of specially requested items can be announced to the user. Now the user can search or browse the stored media and play it out at full rate.

6.2.1 Metadata Requirements

The primary requirement for the metadata component of such a system is simply to make the user's experience of viewing, listening, and reading enjoyable and relevant to his/her needs and desires. To meet this objective requires making or buying a schema, a dictionary, and a set of tools that can evolve and adapt as the needs dictate. The system needs to be capable of describing and tying together all current and anticipated types of content data.

One of the most challenging requirements is low latency. Users place a high value on timely information, so the full metadata lifecycle from creation to consumption has to be able to be performed virtually in real time.

An ideal content preparation system imposes little or no burden on broadcast and media partners, who are at widely varying points in their own metadata deployment for digital production and asset management.

A data broadcasting company not in the business of producing original content does not particularly need a big suite of tools for media production or a system for long-term, deep archival of material.

6.2.2 Standards Efforts

There are an impressive array of works in progress on the part of broadcast-related standards bodies and industry consortia. The SMPTE is developing a metadata dictionary containing hundreds of codeable description keys for audio, video, and other types of data (or essence). The standard provides a good starting point and allows for an official registry for new entries.

MPEG-7, in the drafting stage at this writing, provides a framework and will eventually provide basic tools for describing multimedia material allowing it to be more easily searchable. The standard specifies how audio/visual characteristics, motion of objects, and basic identifiers and encoding parameters can be designed using a large system of schemes and descriptors. MPEG-7 anticipates being able to represent the semantics of a piece of media as well.

The Advanced Authoring Format is a software structure and file format that supports describing and composing complex media objects. It is largely oriented toward production and editing processes, and it incorporates a number of the key elements of SMPTE metadata work.

Another addition to the fray is the TV Anytime Forum, an industry group working toward a specification for describing content available on Personal Video Recorder (PVR) devices.

There are also metadata standards with origins in the Internet community. The most commonly referenced example is Dublin Core Metadata, which was designed as a basic level of description for resources on the Web.

As for tools to create and edit metadata, there are basically two technologies at two ends of the manual/automatic spectrum. The first type evolved from the Web-enabled database paradigm, where free-text fields, radio buttons, and drop-down menus are used to populate fields in a table. The second type is the media logging paradigm, which applies text, audio, and video processing algorithms to live video streams to extract metadata in the form of time-stamped annotations.

A prudent approach includes the following steps:

- Make best use of the best of breed of what is currently available. There is no need to invent a completely new system from the ground up.

- Have the capability to access and incorporate data from existing sources (our content and broadcast partners)

- Participate in and contribute to the ongoing standards where appropriate and available

- Make sure a working system makes it to market in a timely fashion.

6.2.3 Schema Leverage

There is no shortage of standards for uniquely identifying a piece of content. SMPTE has Unique Material Identifiers (UMIDs), and ISO has International Standard Audio/Visual Numbers (ISANs), to name just two. One or more of these can be included in a metadata set as needed to provide an unambiguous reference to the associated content. It is important to note that calling something a "unique identifier" usually means it is an attribute that enforces uniqueness and thus permits an unambiguous reference; this rarely means it is the only such attribute possible for something. For example, a Social Security number, a set of fingerprints, and a California Driver License number are all "unique identifiers" of U.S. citizens licensed to drive in California, but none can be called *the* unique identifier.

One useful and well-developed scheme for describing the format of a piece of data is that used for e-mail. MIME types (Multipurpose Internet Mail Extensions) provide an instant guide to the encoding and interpretation of just about any chunk of data. For news, the International Press Telecommunications Council has developed a useful hierarchical index of subject matter, the Subject Reference System. It has also produced two generations of tagging schemas for news stories: NITF and NewsML. For music, a tagging system known as ID3, originally created for embedding artist and track info on CDs, provides a wide array of information about audio material. The publicly maintained Web directory known as *dmoz* is widely used as a navigation scheme by numerous portal sites, such as Lycos and Google. For geographic information, extensive dictionaries of place names are available free from the U.S. government.

There is also no reason to invent a framework and syntax for metadata (fortunately, because this can be a huge task). For the basic interchange format, XML is becoming the most popular. For the overall schema, the *Resource Development Framework* (RDF) is a good choice. Built upon these underpinnings and with some up-front development, the content creation tools, as well as all of the software for broadcasting, receiving, and end-user interface, speak the same language.

6.2.4 XML and RDF

XML, the eXtensible Markup Language developed by the Web community, is quickly becoming a standard for all kinds of data interchange. It is finding its way into the broadcast world over many routes, including MPEG-7, the MOS interface, and the NewsML standard.

The Resource Description Framework is a foundation for machine understanding of what do with data. If XML provides the ability to describe the syntax of a metadata language, RDF provides the same for its semantics. RDF has a new but stable specification.

Statements

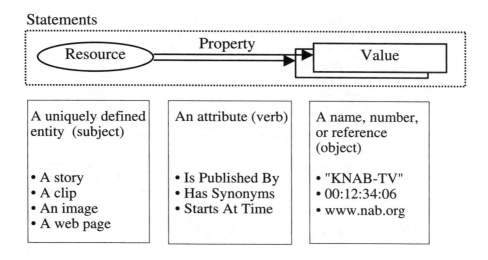

Figure 6.2 The basic RDF model.

In the RDF data model, information is represented in the form of statements, or short sentences (Figure 6.2). This can be represented graphically as a labeled directed graph. Each statement contains a *resource*, or subject; a *property*, which can be thought of as a verb, and a *value*, which can be thought of as the object in the sentence structure.

Statements are lightweight and can be free-standing, which makes it easy to add new metadata to an existing subject.

A resource is any uniquely identifiable entity, such as a particular news story, image, or Web page. A property is really just the name of a tag, comparable to a column in a relational database (RDB) table. Examples are given in Figures 6.3 and 6.4. Values are strings, numbers, or most interestingly, references to other resources. It is this kind of reference that permits rich and complex descriptions to be built.

The following is s a simple example of an RDF description that shows the XML syntax and the underlying model. Suppose a broadcaster has a news clip about Elian Gonzalez with house number 4263. Here's how you say "Clip number 4263 is a story about a person named Elian Gonzalez that takes place in Miami, Florida." The metadata is straightforward and readable (Figure 6.3).

Here is the example again, this time positively identifying the person in the story ("StoryPerson") as a full-fledged resource, enriching what was a

```
<Description about="#4263">
  <StoryPerson value="Elian Gonzalez"/>
  <StoryLocation value="Miami, FL USA"/>
</Description>
```

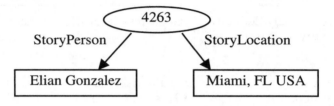

Figure 6.3 Simple RDF example.

simple text name with additional information that turns the name into a unique person description. Furthermore, the Elian resource can be separated out into its own free-standing Description with its own identifier which can be referenced by other relevant clips (Figure 6.4).

A receiving device supplied with this kind of metadata now "knows" more about the story, which allows a user or his/her browser, to search for more intelligently related material.

The schema portion of the RDF spec has a number of useful features for building a metadata model and controlling what kinds of statements can be made—what kinds of things "make sense" to say in RDF.

One main feature is the *class*. A class is just a category or container for resources to fall into. For instance, a data type such as "integer" or "SMPTE time code" can be a class. If certain metadata describes manufactured products, then we might have a class for each category of product, so that we could say "this is a Vehicle", or "this is Apparel". A class can also be a container for the set of values a property can have, such as types of vehicles (sedan, coupe, or roadster). A resource's class membership can also help interpret keywords or other property values. Finally, by using classes and subclasses, conceptual relationships such as "A is a B" and "C is a kind of D" can be expressed. This provides a foundation for inference engines to perform high-level assessments of relevance among pieces of media.

The other main schema tool is the *constraint*. Constraints let the user control the kinds of tags a particular kind of thing can have, and control what

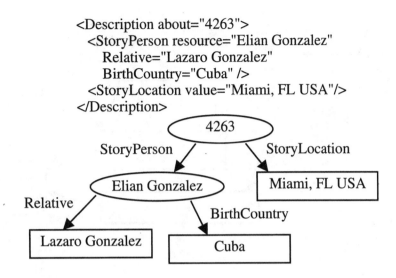

```
<Description about="4263">
  <StoryPerson resource="Elian Gonzalez"
    Relative="Lazaro Gonzalez"
    BirthCountry="Cuba" />
  <StoryLocation value="Miami, FL USA"/>
</Description>
```

Figure 6.4 Enhanced RDF example.

values a particular property can have. Saying what a "kind" is becomes easy to do when we use classes to group things together.

The main reasons for using RDF are:

- The ability to convey knowledge, to go beyond specifying just the way metadata is formatted

- To link entities rapidly, whether the are attached to media or whether they live elsewhere in the world

- To tie all the players and tools together in the end-to-end broadcasting process

After we have a framework for structuring and describing metadata, we are ready to design a tool to prepare it for broadcast. (See Figure 6.5.) Analog and detached media can produce a first level of useful metadata through the logging process. Making the tools XML-based allows, in some cases, original source data in the form of prompter scripts, story titles and authors, and content ratings to be acquired in digital form from a server or newsroom automation system. The narrative metadata obtained in these ways then goes through a keyword extraction and categorization process. Both the categorization and any subsequent editorial process benefit from the presence of a

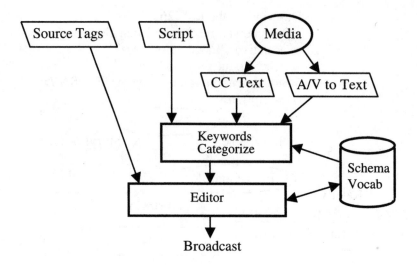

Figure 6.5 Metadata creation process.

controlled vocabulary to accelerate and standardize the tagging process. Editors can be given the capability to modify the vocabulary as required.

At the other end of the broadcast chain are the applications that leverage metadata to enhance the user's experience—browsers, enhanced TV viewers, and the like. The extracted and keyed-in tags make the delivered media and text searchable, browseable, and connected to relevant product information. At this point, some metadata takes on an active role, not just a descriptive role. For example, a content producer may specify that a group of items are intended to be shown together as a package, or that the user must perform an e-commerce transaction to be able to view something. This introduces all the issues of compatibility and capability level matching between the content/ metadata service and the delivery platform.

Ideally, some time in the near future, there will be a single, harmonized standard for metadata, so that content created for broadcast, for the Web, or for recorded media can be repurposed, retargeted, searched and discovered, interrelated, and—above all—understood universally.

6.3   Bibliography

Extensible Markup Language (XML) 1.0, World Wide Web Consortium Recommendation, http://www.w3.org/TR/REC-xml.

Netscape Open Directory Project, http://www.dmoz.org.

"News On Demand," *Communications of the ACM*, vol. 43, no. 2, February 2000.

NewsML, International Press Telecommunications Council, http://www.iptc.org/iptc/.

Resource Description Framework (RDF), World Wide Web Consortium, http://www.w3c.org/RDF/

Sheth, Amit, and Wolfgang Klas (eds.): *Multimedia Data Management*. McGraw-Hill, New York, N.Y., 1996.

SMPTE Metadata Dictionary, http://www.smpte.org/.

TV-Anytime Forum, http://www.tv-anytime.org.

Tvede, Lars, Peter Pircher, and Jens Bodenkamp: *Data Broadcasting: The Technology and the Business*, John Wiley & Sons, New York, N.Y., 1999.

Internet and TV Convergence

Skip Pizzi, Microsoft

7.1   Introduction

Although the Internet and television are viewed on similar video display devices, the difference between them remains profound. The concept of convergence of these mediums is often misunderstood or overplayed. The Internet remains an interactive (2-way) medium carrying largely static content for point-to-point distribution. Meanwhile, television remains a primarily one-way medium with dynamic content distributed in point-to-multipoint fashion.

Interactive television is often considered the true convergence point between these systems, but in fact, it can be more accurately viewed as adapting certain features of each service to create a new hybrid medium. ITV takes the rich, dynamic content of television and adds to it the personalization and responsiveness of the Internet. ITV will likely be seen by consumers as neither TV nor the Internet, but a medium unique unto itself.

The first and most obvious method of merging these mediums is simply putting Internet content on a television screen, or conversely, viewing TV content on a computer. Neither of these processes is as simple as it might seem, but the latter is generally more easily accomplished. With the proper hardware, television programs can be fairly successfully viewed on a personal computer screen. On the other hand, putting a Web page from a computer browser output directly onto a TV screen is generally not a satisfying experience. This is because Web content is typically viewed on a computer by a single user positioned at close proximity to the screen, while television is usually watched at far greater distances and often by groups of viewers. This distinction is referred to as the "one-foot vs ten-foot experience" or "lean forward vs. lean back" usage. This implies that when viewing the typical Web page on a TV screen, fonts and graphics are generally too small to be comfortably viewed. Also, the selection of hyperlinks can be difficult with an infrared remote control.

ITV systems have dealt with this issue in different ways. Some systems transcode the Web page content in a specialized server so that it displays more appropriately on a TV screen. Other systems perform this transcoding in the client receiver, or with a combination of client and server-based steps. These transcodings typically involve font substitution to a larger typeface, while retaining as much of the look and feel of the original page as possible. The overall success of the process varies widely. Some systems do a better job than others, and some Web pages lend themselves to such transcoding more readily than others.

In addition, because a mouse is generally not used with ITV receivers, Web-page links are converted to another form of display. Common practice to date has been to highlight each link on the screen one at a time, using the remote control navigation keys to sequence among the links. For example, as a user pushes the "down arrow" key on the ITV remote, the next link down the page will be highlighted. The user steps through the links on the page using the navigation keys until a link that the user wishes to follow is highlighted. Then the user presses a "Go," "OK" or "Enter" key on the remote, and the display switches to the linked page, once again transcoded for TV display.

The ultimate solution to this problem involves the design of Web pages customized for TV display by the content author. While this is rarely done at present, it is likely to become standard practice as the ITV medium takes hold.

7.2 ATVEF

The *Advanced Television Enhancement Forum* (ATVEF) was formed in 1997 by a consortium of 14 leading companies in the television and computing industries. This group developed a public, worldwide specification for creating and delivering interactive TV content, based largely on existing content standards from the Internet world. Over 100 other companies have become "adopters" (i.e., licensees) of the format, which allows them to incorporate the standard in products or programs they produce. The ATVEF Specification v 1.1, r26 was finalized and published in 1999, after which the organization sunset itself as an active group. Interested parties are still able to become adopters of the format, however, through the ATVEF licensing office.

A successor group of sorts, the ATV Forum, was formed in 2000, with the primary intention of marketing and expanding the influence of the ATVEF specification. This group includes many of the original founders of ATVEF, but it does not intend to develop further technical specifications. That work has largely been taken up by the SMPTE, under its *Declarative Data Essence* (DDE) specialist group. At press time, SMPTE was preparing a DDE-1 specification, which would extend ATVEF v1.1 into an internationally recognized format.

The ATVEF specification's fundamental premise is the COPE principle: "Create Once, Play Everywhere." The format was therefore written with the intention of allowing ITV authors to write content once and deliver it to a variety of platforms. These platforms include intelligent TV receivers (analog and digital), set-top boxes, and PCs. The format accommodates both one-way and two-way modes of operation, and offers methods of interoperability with all international television formats.

7.2.1 The Specification

The ATVEF Specification v1.1, and its successor SMPTE DDE-1, contain the following primary components:

- Content specifications to establish minimum requirements for intelligent receivers

- Delivery recommendations for the transport of enhanced TV content over various distribution formats

- Bindings for content to the respective delivery formats

A key design point of the specification involved the use of existing content and delivery standards whenever possible. The creation of new methods was done only when necessary to conform to the unique requirements of a broadcast ITV architecture. Therefore, the ATVEF specification references the full existing specifications for the following:

- Hypertext Markup Language (HTML 4.0)

- ECMAScript (ECMA 262)

- Document Object Model (DOM 0)

- Cascading Style Sheets (CSS 1)

- Various other media types

The specification does not set an upper limit on such content, but provides a nominal capability set for content developers.

A second primary design goal of the format was the provision of a single solution that would run on a wide variety of delivery networks. With the proper bindings, it is possible to carry ATVEF-compliant content on either analog or digital television systems, as well as systems that do not support video at all. This includes transmission over terrestrial broadcast, cable and DTH satellite systems, as well as over the Internet. Analog TV broadcast systems would typically use the vertical blanking interval for such content, and digital systems could use a variety of delivery methods including data carousels, data transport streams, IP over video, and other applications.

The format also acknowledges the "food chain" of broadcast delivery, so it is designed to allow ATVEF-compliant content to bridge between networks, such as in the case of data on an analog terrestrial broadcast bridging to a digital cable system. This functionality was made possible via the implementation of transport-independent content and the use of IP as the reference binding, by way of the IP bindings already defined for various television systems.

The ATVEF specification does not specify the bindings for each television delivery system, however, preferring to leave that to the standards bodies that controlled each of those formats. It sets as a default the IP-over-TV bindings that each TV broadcast format defines, using the IP Multicast protocol for data transport to the delivery system. The ATVEF specification also includes appendices for the Unidirectional Hypertext Transfer protocol (UHTTP) for optimized IP delivery in a one-way, point-to-multipoint broadcast environment, as well as an example of a transport specific binding set for the NTSC television format.

7.2.2 Content Types

A significant difference between the on-line and television broadcast worlds is the need to accommodate one-way reception of data on at least some interactive TV receivers. This implies that content creators cannot customize their data for different receivers or browsers as they do in the two-way environment of HTTP. One way the ATVEF specification addresses this problem is by declaring a base profile of Multimedia Internet Mail Extension

(MIME) file types that must be supported in all ATVEF-compliant receivers. These files types include the following:

- Text/html (HTML 4.0)
- Text/plain
- Text/css (CSS1 only)
- Image/png
- Image/jpg
- Audio/basic

In addition, ATVEF recommends, but does not require, support for the following MIME file types:

- Image/gif
- Audio/wav

The ATVEF specification also accommodates the integration of TV pictures into Web pages. For this, the specifications defines the *tv:URL*, which can be used to identify a specific broadcast channel as a source for a video image to be placed on a Web page. Similarly, the specification's *lid:protocol* supports a locally stored video source. In either of these operations, the video image is scaled and placed in a window the Web page using the HTML <OBJECT> and <IMG> tags. The image can also be placed as a background in a table cell or the entire page, onto which the data of a Web page can be overlaid.

7.2.3 Transport

The ATVEF specification defines two different transport modes, entitled *Transport A* and *Transport B*. Transport A refers to situations where a data-return path (or "back-channel") from the user is available, while Transport B covers one-way broadcast data. Transport A content generally includes only URIs ("links") and triggers, with actual ITV content resources accessed from the Web or other on-line sources. In other words, ATVEF Transport A content directs the ITV receiver to other on-line locations to gather resources (in many cases by caching them in advance), and activate them when the appropriate trigger is received.

ATVEF Transport B, on the other hand, includes all the ITV content resources in the broadcast stream. This implies that significantly greater bandwidth in the broadcast channel may be required for Transport B, but no backchannel is required.

The standard also defines a method for a single ITV program to support both Transport A and B simultaneously. In this case, a program carries enhancement data intended for bidirectional IP-type connections as well as for one-way, broadcast-only receivers, which would allow the content to accommodate a wide range of broadcast delivery systems and receiver types. ATVEF receivers can choose between these formats when both are encountered.

Note that the ATVEF specification only specifies a *content* format. The transport-related elements of the ATVEF specification are presented simply as examples of how that content might be successfully transmitted. Because the ITV industry was in its earliest days when ATVEF was formed, the specification deals with transport issues at some depth, although it takes pains not to *specify* particular transport mechanisms.

7.2.4 Triggers and Announcements

Triggers are real-time events delivered for the enhanced TV program that determine the running time of various data elements, synchronized with reference to the television program they accompany. The ATVEF specification includes details for specifying these "markers" in ITV content.

Various receiver implementations can unilaterally determine how users will enable or disable display of such enhanced TV content. For example, trigger arrival can be used as a signal to the receiver, which can then notify the user of enhanced content availability. Alternatively, the receiver can be set to automatically display such content upon arrival, or to ignore it completely.

In the one-way broadcast mode (Transport type B), *announcement* data is used to initiate or offer enhancements to the TV program. An announcement specifies the location of both the *resource stream* (i.e., the files that provide content) and the *trigger stream* for an enhancement. These files will likely have been cached in the local memory, for which the ATVEF specification requires a minimum of 1 MB be provided in the receiver.

Announcements also provide information about the enhancement's language, its start and stop times, its bandwidth, and its peak storage size needed for incoming resources, along with other data. Among the latter, an option exists for the identification of another broadcast channel for cases in which ATVEF-compliant content is sent separately from the audio/video TV program. The receiver must be able to start receiving data using only the description broadcast in the announcement.

The data format of an announcement follows the ATVEF specification's *Session Description Protocol* (SDP), which is preceded by a *Session Announcement Protocol* (SAP, not to be confused with the *Second Audio Program* of multichannel television audio formats, which shares the same acronym).

An additional time-related element of the ATVEF specification involves how to cancel previously delivered enhancement data and remove it from the local memory after it has expired (for example, after the broadcast program that the interactive data accompanies has concluded).

7.2.5 Bindings

Bindings in the ATVEF specification define how ATVEF-compliant content will run on a given delivery system. The binding may support either or both Transport types A and B. It is necessary that a standard ATVEF binding exist for each transmission format so receivers and broadcast tools can be developed independently but remain interoperable. The binding provides the "glue" between the delivery format's specification and the ATVEF specification, in cases where the network specification does not contain all the necessary information for such data carriage.

Because carriage of IP is already included in the specifications of many media-transmission systems, ATVEF defines its binding to IP as a reference binding, as noted previously. This implies that one way to build an ATVEF binding for a particular network is to simply define how IP runs on that network when associated with a particular video program.

The ATVEF specification also includes a model of how to build a binding specific to a network standard, using NTSC as an example. ATVEF-compliant data bytes are encoded into the vertical blanking interval (VBI) of NTSC fields. For Transport A, Line 21 of the VBI is used, following the Text-2 (T2) service specified in EIA 608 (recommended practices for Line 21 services). The data encoding follows EIA 746A, which describes how to send URLs and related data on Line 21 without interfering with closed captioning and other data that share this VBI line.

For Transport B in the example binding, the aforementioned IP reference binding is applied, using the Internet Engineering Task Force's (IETF) IP-over-VBI recommendation. The NABTS encoding format is used (as is standard practice for NTSC VBI data), with ATVEF data streams constrained to packet addresses 0x4B0 through 0x4BF.

7.2.6 UHTTP

An Appendix to the ATVEF specification defines a one-way, IP-multicast based, resource transfer protocol called the *Unidirectional Hypertext Transfer Protocol* (UHTTP). It is designed to efficiently deliver resource data in a one-way broadcast-only environment. It can be used to carry IP Multicast data on an analog TV signal's vertical blanking interval (IPVBI), or in the MPEG-2 DTV stream, as well as other unidirectional transport systems.

This IP Multicast stream can carry Web pages and their related resources (including scripts) alongside the related TV signal. A session announcement broadcast by the TV service tells the receiver which IP Multicast address and port to monitor for the associated data.

7.3   TV Anytime

In 1999, an international industry consortium was formed to set standards for the emerging *personal video recorder* (PVR) industry. The consortium, entitled the TV Anytime Forum (TVAF), set as its goals the establishment of primarily content-related standards, which would allow global interoperability of digital media recording equipment. (See Figure 7.1.)

To accommodate this goal, TVAF is developing (at this writing) three specifications and two recommendations. The specifications cover the areas of content referencing, metadata, and content rights management; while the recommendations deal with business models and system design. TVAF's began issuing its first draft specifications in mid-2000.

7.3.1 Specifications

The TVAF content referencing specification deals with TV program identification and location data. Content that includes this ID data, and which is delivered on a TVAF-compatible network, will be locatable by any TVAF-compliant end-user recording device or system. Ideally, this includes both scheduled broadcast programs and media content stored on Internet servers. The TVAF specifications also support personal media storage that is physi-

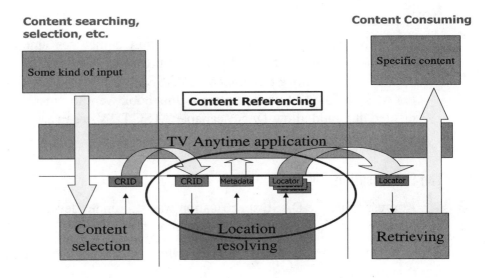

Figure 7.1 Overview of the TV Anytime Forum realm. (*Courtesy of TV Anytime Forum.*)

cally located on a PVR at the user's premises as well as remotely stored data in private shares of on-line servers.

The TVAF metadata specification will standardize a language for descriptive data about a TV program, by which a TVAF-compliant device or system could locate content via intelligent searching. The Forum's content rights management specification would accommodate copyright protection and transactions for downloading and viewing of premium content.

In addition to these normative specifications, TVAF is developing some informative documents that are intended to help broadcasters and equipment manufacturers pursue a coherent vision in their development of PVR-based products and services. These include a description of potential business models and opportunities that the PVR (or PDR—Personal Digital Recorder, as TVAF calls it) permits, as well as a technical overview of end-to-end system designs that support TVAF-compatible content.

All the TVAF specifications accommodate delivery environments of either one-way or two-way connectivity, as well as those with or without conditional access and content security features.

7.4   DASE

The Advanced Television Systems Committee includes among its many components a specification for ITV applications. This is called the *DTV Applications Software Environment* (DASE), which was being developed by the ATSC's T3/S17 Subcommittee as this book went to press. When implemented, it would allow DASE-capable ATSC DTV receivers to decode and display standardized ITV content.

7.4.1 The Specification

At press time, the DASE specification was still under development. Current plans for the platform include two main parts:

- A *Declarative Application Environment* (DAE) for the display of Web-style content

- A *Procedural Application Environment* (PAE) for the execution of Java-based procedural content

The DAE is a browser-like environment based on modularized XHTML, ECMAScript, and DASE-specific forms of the W3C's Cascading Style Sheets (CSS) and the Document Object Model (DOM). (The W3C standards were developed for Web content, and their DASE variants address the specifics of their application to ITV.) The DASE DAE includes a Java Virtual Machine (JVM) and associated ATSC-specific APIs.

An open issue at press time was the *profiling* (i.e., functionality tiering) of DASE receivers. There may be DASE receivers that support only declarative content and others that include procedural content capability. It was also unsettled whether and how ATVEF content would be accommodated in DASE.

7.5   PVRs

An important development in the television environment, and one that has strong resonance for ITV, is the introduction of the personal video recorder (PVR). Because these devices allow personalization of programming, they are attractive to consumers and offer a fundamental form of interactivity with the content environment for consumers.

These devices allow a substantial amount of television programming to be stored on a hard disk and played back in non-linear fashion. Because a non-linear storage device (a hard drive) is being used, these devices can begin to replay a program while it is still being recorded. This allows a powerful feature of the PVR called *live pause*, by which the device is used as a buffer for live programming. This permits users to pause a program that they are watching as it is currently being broadcast, and return to it a few minutes later without interruption.

PVRs also offer a range program selection processes from simple searching of electronic program guides to intelligent searching for keywords in program titles, subjects, content, talent names, and so on. It is likely that future PVRs will incorporate a browsing ability to seek and download stored content from on-demand libraries, as well as access broadcast program schedules. Many of today's researchers forecast that a significant portion of future television viewing will be from PVRs rather than from real-time broadcast sources.

Some current PVRs, and most future units, also store ITV elements of a program so they can be selectively accessed upon later playback as well. A PVR that can access both broadcast channels (via terrestrial, satellite, or cable) plus the Internet via broadband connectivity will serve as a powerful media collection tool. When interfaced to a home network system, such a storage device could act as the home media server. It is easy to see why the PVR is considered such an important development in the TV environment.

While today's PVRs are standalone devices, those of the future may be part of a PC, a home server, or even a remote storage system in which the actual physical storage is on a server at some service provider's location, with unique personal access by the user. In this respect, the PVR is essentially a user interface that accesses storage locally or on various networks, with the actual storage location of a particular program being transparent to the user. The ultimate vision of this process is the creation of virtual channels, by which the users's PVR build the program stream that the user views from various stored sources.

Numerous business models have been proposed for PVR-based systems. The simplest involves subscription to EPG data (as in current services offered by TiVo, Replay Networks, and WebTV). Future scenarios involve VOD services, selective commercial skipping or replacement, intelligent global searching, automatic highlighting, sharing and transferring of stored videos or clips, rental of off-line storage space, and much more.

Content Distribution

Chandy Nilakantan, SkyStream Networks

8.1 Introduction

The Internet is not just ubiquitous, its use for conducting commerce is becoming pervasive. Great strides have been made in the areas of providing widespread connectivity, increased capacity for moving richer content and adequate security for maintaining privacy. One of the largest remaining challenges is scalability—particularly for distributing and delivering mixed media content to large numbers of locations and users in a predictable and scalable manner. The Internet has proven to be a great transaction network but falls woefully short when it comes to delivering services to even moderately sized communities or subscriber bases. On the other hand, present-day broadcast infrastructures, satellite, cable, and terrestrial networks possess built-in scalability and have the means to deliver broadband services to communities of any size, spread over any geography, very effectively.

8.1.1 Trends in Internet Content

The growth of data traffic over the Internet is driven by a number of factors, including:

- The increase in the number of consumers accessing Internet on a more frequent basis

- Increase in the number of businesses using the Internet not only to reach customers, but also to conduct business transactions with partners, vendors, and suppliers

- Proliferation of new wired and wireless Internet enabled devices

- Increase in the availability and use of broadband Internet access services

- Proliferation of increasingly complex multimedia-rich content

- Emergence of new Internet-based applications designed to simultaneously reach large audiences

Content on the Internet is evolving rapidly from simple, static Web pages to complex graphics and audio/video streams. Delivery of this new class of content requires significantly more bandwidth. In addition, while the Internet has historically been used for one-to-one exchanges between a user and a host server, new applications are emerging, such as online learning and live webcasts, that require rich content to reach millions of users simultaneously. These applications require *one-to-many exchanges,* and are placing great demands on the Internet infrastructure in terms of bandwidth, scalability, and predictability of content delivery.

8.2   Limitations of Internet Infrastructure

The current Internet infrastructure is not designed to efficiently handle the delivery of rich, multimedia content to a large number of simultaneous users. There are two fundamental reasons for this:

- The Internet is a *point-to-point* network. The Internet is built to handle movement of data from one point to another. When the same data needs to be sent to several locations, additional copies of the same data are sent, separately, on each path that connects the sender to the individual receivers. As the number of simultaneous receiver locations increases to millions, the Internet infrastructure hits a scalability wall.

- The Internet is a *best-efforts* network. The path between each user's personal computer and the computer that holds the requested content is a series of interconnected computer networks. Data moves through these networks in a hop-by-hop manner. Not all networks have the same capacity to carry data and not all devices that interconnect these networks have the same capacity to process the data at similar speeds. If any of the intermediate hops experiences congestion, as is the case when a large number of users try to access the same content, some of the data packets can be dropped arbitrarily or delayed. These errors introduce a degree of unpredictability and unreliability in content delivery. When the content includes audio and/or video streams, this loss or delay of data packets can severely impact the listening and viewing quality.

In recent years, large investments have been made to improve the physical infrastructure of the Internet and to increase the amount of available bandwidth. Advances in networking technologies such as gigabit and terabit-speed routing, optical networking, broadband access, Web caching, and content replication have delivered solutions that greatly increase the Internet's capacity to carry high bandwidth content from point-to-point. For example:

- *Last mile access*: Users enjoy rich content by using access technologies like DSL and cable modems, which have greatly advanced the carrying capacity of the "last mile."

- *High speed routers*: Advances in ASIC technology and high-performance software enable routers to handle complex network topologies and to process IP packets at gigabit and terabit speeds. ISPs can deploy these routers and provision faster links to the core.

- *Optical networking*: Wave division multiplexing (WDM) and *dense wave division multiplexing* (DWDM) have increased the available bandwidth in the core of the Internet by several orders of magnitude.

In spite of increased last mile bandwidth—faster routers moving packets more efficiently and quickly through the network and a high capacity core—there continues to be a lack of predictability of response time for transactions on the Web. When the same content is being requested by a large number of users, the congestion point is no longer the network—it is the data center that serves the content. New strategies are now employed to speed up the overall response time of the Internet, by moving the content closer to the edge of the network and therefore requiring fewer hops from the user.

- *Web caching*. Web caching can be used to pre-fetch commonly requested content and have it stored in locations that are closer to the user, thus eliminating the need for every transaction to be serviced by the data centers. Web caching can be used to effectively decongest data centers, but it only works when the content being served is *static*. In other words, if the content changes frequently, then the caches have to be replenished frequently as well, and this can result in stale caches and unpredictable delays.

- *Content replication*. Content replication is an effective method for improving response time to transactions, particularly when one can predict which content will be in demand and at what locations. Content that changes frequently, such as real-time stock updates, news highlights, and live Web events can be replicated to many servers at locations close to users. Content serving occurs locally, and therefore delivery is more predictable.

8.2.1 Emergence of Content Distribution Networks

Content distribution networks (CDNs) aim to help Web site owners deliver their content to consumers and businesses with greater performance and higher reliability. The underlying method used by content distribution networks is replication; by making copies of the content available closer to where the users are, the overall Web access response time is reduced greatly. There are two predominant modes of operation for CDNs:

- Data center based operation, where the content distribution network provider manages the service out of private facilities or co-located facilities. Typically, in this mode, content providers or website owners are the direct beneficiaries of the service.

- ISP *point-of-presence* based operation, where the content distribution network provider's service is oriented towards improving the performance of the Web cache located at the ISP. In this mode, the ISP is the direct beneficiary of the service.

The network infrastructures used by these service providers to move the content to the edge locations vary greatly, from shared to dedicated, and from point-to-point to point-to-multipoint.

All of the content distribution networks in operation today are designed for specific types of content, and specific types of services that are delivered based on that content. For example, some of these networks are intended for streaming media services only and therefore transport streaming audio and video. Others are for speeding up Web response and transporting Web objects. Still others are for delivering broadcast television and radio services and transporting digital audio and video programs. The type of services associated with the content they carry dictates the underlying network architectures employed by these networks.

But, clearly, content types are converging rapidly. Television broadcast programs are showing up on Web sites and Web-based content is being integrated into TV programs. But the networks that transport this converged content are ill prepared for the quality of service required to meet the demands of the applications that employ this content.

8.3   Broadcast Internet

Unlike the Internet, broadcast networks have been optimized for the transmission of rich content to large numbers of users in a predictable, reliable, and scalable manner. The advantages they bring to the infrastructure are many:

- Broadcast networks are designed to carry rich, multimedia content. Traditional broadcast networks—including television, cable, and satellite networks—have been explicitly designed to deliver high-quality, synchronized audio and video content to a large population of listeners and viewers. In addition, over the last five years many of these networks have migrated from analog to digital transmission systems, thus greatly enhancing their ability to carry new types of digital content, including Internet content.

- Broadcast networks are inherently scalable. By virtue of their point-to-multipoint transmission capability, broadcast networks are inherently scalable. It takes no more resources, bandwidth or other provisions, to send content to a million locations as it does to one, as long all of the receiving locations are within the transmission footprint of the broadcast network. In contrast, with the traditional Internet, each location that is targeted to receive the content will add to the overall resources required to complete the transmission.

- Broadcast networks offer predictable performance. Again, by virtue of the point-to-multipoint nature of transmission on broadcast networks, there are no variances in the propagation delay of data throughout the network, regardless of where a receiver is located. This inherent capability assures a uniform experience to all users within the broadcast network.

By enabling broadcast networks to connect with the Internet, the *Broadcast Internet* offers a cost-effective, reliable, and seamless path for delivering multimedia rich content to large numbers of users and service providers simultaneously. Pan-continent broadcast networks, such as digital satellite systems, can be used to distribute content to a very large, and potentially highly dispersed, set of locations. Other types of broadcast networks with smaller footprints, such as digital TV networks, can be used for local content distribution. The content is received and cached locally at these locations, which will—in turn—serve the content to users that connect either directly to that location or through a service provider that uses the location for content serving. The content is served to the users using the traditional mechanisms employed on the Internet. For example, instead of the ISPs replenishing their local caches on an on-demand basis, they can adopt the strategy of pre-filling their caches based on analysis of the type of content most frequently requested by the communities they serve. Content that is dynamic in nature, such as audio/video streaming and stock tickers, will be updated on the local caches at the appropriate frequency, keeping the cached content current and providing better overall quality of service to the users. Figure 8.1 illustrates the application of this principle to a cable-based delivery system. Figure 8.2 expands the concept to a variety of distribution media.

The major design considerations for building Broadcast Internet-ready broadcast facilities are:

- Multi-vendor interoperability. Any equipment that is added to the head-end to incorporate Internet data must work seamlessly with all existing equipment and must fit within the operational framework used by the facility. This requires any new equipment to support standard interfaces and protocols, and follow the guidelines set by standards organizations that cover both the digital video and IP domains.

- Bandwidth optimization. Managing the available bandwidth in optimal fashion is a very important design consideration. There is considerable variability in the use of bandwidth by compression and multiplexing equipment, caused either by the inherent inefficiencies of the equipment, the nature of the content, or the service mix. The head-end architecture must provide the ability to capitalize on this variability to maximize the use of the bandwidth in an opportunistic manner. There are two strategies that are commonly applied for allocating bandwidth to data services in an environment that includes legacy video services as well: 1) *pre-allocated bandwidth*, where the data services are treated just like another "channel" or "bundle of channels" and the bandwidth is reserved for the exclusive use of these services; 2) *opportunistic bandwidth*, where the data services are treated as "trickle services", where there are no specific time of delivery requirements for the service. In this case, bandwidth can be optimized by not pre-allocating the bandwidth for the data services and taking

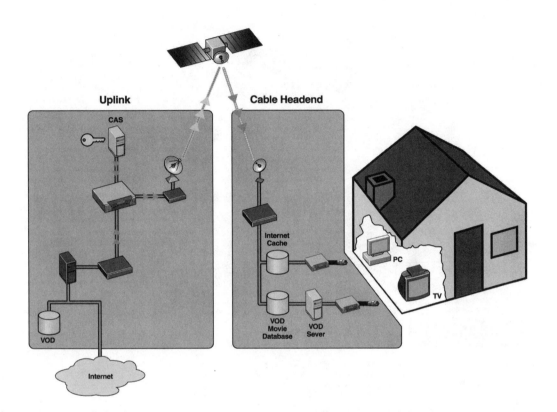

Figure 8.1 A high-bandwidth data distribution system incorporating cable TV distribution to end-users.

advantage of the variability in the bandwidth use by the legacy video programs to "fill the blanks" with data.

- Flexibility. The nature and type of Internet services that will be required to be supported in any head-end is likely to change rapidly, as is evidenced by the rapid rate of technological and application advancement in the Internet in general. The head-end architecture must easily adapt to new applications requirements without having to undergo major design changes.

Currently, applications that incorporate data can be classified as following:

- Pure data services. These services use Internet or private data as the sole content. There are no legacy, television-oriented digital audio/video ele-

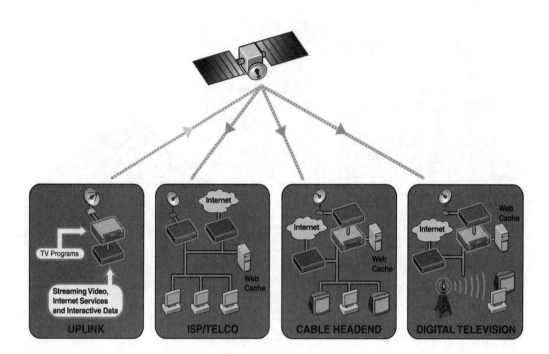

Figure 8.2 High-speed data delivery using a variety of last-mile systems.

ments in the content. This type of content is typically targeted towards personal computers and integrated set-top boxes that have an operating environment similar to personal computers and are enabled for interactive services on the television. The content preparation, packaging, scheduling, and delivery systems used for these services can be independent of those used for the legacy video services.

• Loosely co-related data services. These services use Internet or private data to enhance or extend a legacy, television-oriented program, or to provide a parallel co-related content stream that is independent of the video stream, but scheduled for play-out simultaneously. The parallel content stream can be viewed in real time or stored for later viewing. The content preparation, packaging, scheduling, and delivery systems used for these services must have the ability to interface and interact with those used for the legacy video services.

- Enhanced TV. These services use Internet or private data to enhance a legacy, television-oriented program in a highly synchronized manner. The synchronization can be either presentation time-based or frame content-based. The content preparation, packaging, scheduling, and delivery systems for these services must be fully integrated into those used for legacy video services.

8.3.1 DTV Content Distribution Models

There are three different models that can be employed for content distribution over DTV:

- **End-to-end DTV data broadcasting**. In this model, the service provider creates services by aggregating content from both the TV domain and the Internet domain, distributes these services over a satellite or land-based network to television stations in the targeted regions. The television stations—either in real time or in a *store-and-forward* manner—then deliver the services. The customers for such services would be consumers.

- **Data distribution to Internet service providers.** This model is suitable for local distribution of content, particularly multicast services and streaming media, to Internet service providers that then deliver that content to consumers over traditional land-based last-mile connections. This model can be very effective for simulcast services—programs that are viewable on the TV as well as on a PC, with enhancements specific to the PC.

- **DTV as last mile**. In this model, the DTV spectrum is used as a last mile in a content distribution network. Content providers lease or acquire on demand last mile bandwidth for delivering Internet based services.

8.4 Summary

Content providers are shifting their focus rapidly toward high quality, reliable delivery of new types of content that is a blend of traditional television fare and the Web. The networking industry is now looking seriously at the enormous potential of broadcast networks, including DTV. As the need to move large amounts of content in a scalable and efficient manner increases each day, hybrid networks that combine broadcast and broadband become extremely valuable. The television industry should view the DTV opportunity not merely as high-quality TV images or fat pipes for data, but rather as an opportunity to revolutionize the networking industry.

PC-Based Receivers

Kishore Manghnani, Teralogic

9.1   Introduction

The personal computer (PC) can initially play a major role in stimulating the growth of DTV. A huge installed base (over 350 million units), open standards, and low cost makes the PC an ideal platform for deploying DTV in consumer homes during the initial ramp- up phase. The PC monitor, with its million-pixel resolution, is already capable of displaying high-definition TV (HDTV) images. Additionally, the deployment cost is significantly lower on a PC platform. An HDTV PC add-in card can cost considerably less than an HDTV set, making it attractive to consumers.

Many consumers are currently using analog TV cards to watch TV on a PC screen. This represents a sizable market segment that has already combined the TV and PC experiences. The market for PC analog TV cards has grown steadily over the last few years with shipments expected to reach 4 million units annually as this book went to press. With its rich viewing experience and enhanced interactive data services, a DTVPC card is a more compelling solution than current analog TV tuner cards for PCs. The DTV experience is significantly enhanced compared with analog TV, as "cinematic" entertainment is combined with the real-time interactivity of the PC. A potential analog TV card buyer would be inclined to spend a little more for a combination card that supports both analog and DTV broadcasts in order to avoid the risk of quick obsolescence. Also, the consumer can sample the HDTV experience on a PC monitor before committing to a sizeable investment in an HDTV set for the family room. Broadcasters will be encouraged to create more DTV content if the installed base of DTVPC cards grows, which will help the DTV industry as a whole.

9.1.1 A View of the Convergence

Only a couple years ago, there was a clear differentiation between appliances in the home. A television displayed pictures and sound from broadcast sources, the stereo system played music from CDs and other sources, the phone, e-mail, and video games were facilitated by different distinct appliances. Today, however, we see these and other devices evolving into single units. The video game not only entertains the family—some are now capable of e-mail and browsing over the Internet. The home PC is also evolving. Not only do we send e-mail and write letters on our PC's, we now can attach sophisticated sound systems to replace the stereo system, and have incredible gaming and other convergence functions running on the PC.

Intel, Microsoft, Compaq, and others have been working aggressively to bring DTV technology to the PC platform. The PC industry has a huge market of more than 100 million units shipped annually. This offers a lucrative opportunity for cleverly designed DTV decoder solutions. The decoder architecture should facilitate design of PC/DTV add-in cards to broaden the reach of DTV. PCI-based DTV decoder cards will have a significant advantage in penetrating the PC market, as PC/DTV boards will be easier to design and upgrade. Compliance with the Windows 2000 and future operating systems is absolutely necessary for PC platforms.

The DTVPC cards continue this trend of converging existing home appliances into a single entity. We can now display high-resolution video and CD quality audio of digital broadcasts on the PC screen, or use the card to bring digital broadcasts to an existing television.

9.2 Features Required in a DTVPC Solution

There are specific requirements a tuner card must have in order to be suitable for DTV/PC convergence, including:

- Dual-input tuners capable of supporting multiple DTV formats must be used.

- Because the bulk of TV programming will be analog for the next few years, a DTV card must be able to support existing analog TV broadcasts.

- The DTV card must support multi-sync monitors and common PC display resolutions such as 1024×768, 800×600, and 640×480. In the U.S., it must be able to decode and output all 18 ATSC formats, so it can display

Table 9.1 The Basic 18 ATSC Display Formats

Vertical lines	Pixels	Aspect Ratio	Picture Rate
1080	1920	16:9	60I, 30P, 24P
720	1280	16:9	60I, 30P, 24P
480	704	16:9 and 4:3	60P, 60I, 30P, 24P
480	640	16:9	60P,60I, 30P, 24P

any DTV broadcast, regardless of the transmission format of any particular station (Table 9.1).

9.2.1 Mass Market Acceptance

While the actual price points for DTVPC solutions will be defined by market conditions, the first implementations of hardware DTVPC solutions are in the neighborhood of $300. This price point is considerably less than the $2000 to $5000 that is required to purchase a TV/set top box solution with the same capabilities of a DTVPC solution. (The user must, of course, already own a PC.) This is one of the most attractive aspects of the DTVPC card. The end-user can, for a modest investment, enjoy a rich viewing experience and evaluate the remarkable features and quality of HDTV with no more expenditure than the price of the add-in card and antenna for reception of the broadcast signals.

9.2.2 Support of PC Screen Formats

Unlike the existing NTSC analog standard that only supports one mode of operation, the ATSC digital broadcast standard defines multiple modes. These features allow the broadcaster to specify which mode they wish to use based on their own criteria. A broadcaster may wish to send a standard definition picture to the viewer to limit the bandwidth required to send the picture or they may wish to send a full high-definition picture for optimum quality. This means the viewer must have a television capable of displaying these multiple formats.

One of the more convenient aspects of the DTVPC solution is that the PC monitor is ideally suited for display of HDTV. The PC monitor is capable of responding to a wide range of screen formats and resolution, including the various screen formats and resolutions required to support HDTV. This means that most consumers will not have to by a new monitor to display HDTV.

For example, consider the HDTV screen display formats. A 720P DTV stream has a screen format of 1280×720 @ 60Hz. Most PC monitors easily work within these parameters for the screen display. The same is true for the other screen formats specified by the ATSC DTV standard. Most PC monitors support formats such as 720P, 480P and 480I, and many are capable of 1080I format.

9.2.3 Feature Set in a DTVPC Solution

All of the features that we are currently seeing in set top box implementations are possible in DTV-PC solution, again at a lower cost because of the existing hardware of the PC. For example, the ability to pause a live broadcast while the viewer is otherwise occupied is implemented with a *time shift* function. In time shift recording, the system needs a fast access large storage medium to save the program for later display. Current set top boxes may include a hard drive in the system for this purpose. This is easily implemented in the DTVPC system because of the existing presence of the system hard drive.

Some of the primary functions and their benefits are:

- **Multicast.** Some broadcasters are already planning to multicast two or more choices of programming. This is particularly useful for sporting events where multiple games are carried by the same broadcaster. Instead of the broadcaster selecting the game that will be shown in a particular region of the country, the viewer will have the opportunity to select the game they wish to see, or to see all of them at the same time.

- **Record/playback**. These functions are identical to the like-named functions of VCRs or tape recorders where the viewer has the opportunity to save a broadcast transmission for viewing at a later time. Instead of using a tape to store the audio and video content, the DTVPC will use the hard drive of the PC to store the recorded information.

- **Pause/time shift**. This function is similar to the pause function of a VCR or recorder, except in this case the viewer has the opportunity to pause a live broadcast and then resume the playback while the live broadcast is still underway. The DTVPC solution is capable of this type of functionality because of its high-speed peripherals, in this case the internal hard drive. Most hard drives can handle data fast enough to be able to record and playback simultaneously.

- **Data broadcasting**. Technically a DTV broadcast is a simple continuous stream of data. Imagine a very fast network connection sending pictures, sounds, multimedia games, and illustrated articles—all related to the television program the viewer is watching. The consumer can still passively watch TV, but can also customize the experience and make it their own.

- **Interactive TV**. Digital television is heading toward a convergence with computers—and students, sports fans, news junkies, and anyone with an interest in anything will get more out of television. The market potential for interactive data services on TV is huge, with applications ranging from on-line shopping and video-on-demand to viewer participation in real-time sporting events. Real-time interactivity allows the viewer to purchase music CDs during an MTV broadcast or play along with the contestants in a game show.

DTVPC cards represent an ideal element in the interactive TV market. This is due to the fact that most PCs already have the most important function required to implement interactive TV—a mechanism for the viewer to provide feedback. The Internet connection found on most of today's PCs is the medium by which viewers will communicate back to the service provider with information ranging from response to game show questions to political polling data.

9.3 Practical Implementation Considerations

There are three basic ways to bring the DTV experience to the PC platform:

- Software DTV decoding
- Hardware assisted software DTV decoding
- Hardware DTV decoding

Each of these approaches will be examined in the following sections.

9.3.1 Software DTV Decoding

The software decoding approach forces all of the processing work on the system CPU. There are three components in the software-based solution: a

Table 9.2 Software-Only DTV Decoding Workload

DTV Hardware	Over-the-air reception VSB demodulation
System CPU	ATSC demultiplex Full MPEG-2 decoding AC3 decoding Applications overhead
Graphics Hardware	Screen display

receiver card, a high-powered CPU, and the graphics chip. The receiver card, consisting of a DTV tuner and a vestigial sideband (VSB) demodulator, receives the DTV broadcast signal and passes it to the CPU for further processing. The CPU demultiplexes the transport signal and decodes the DTV data using software algorithms.

The software approach strains the CPU tremendously and takes up most of the bandwidth for 1080I resolution. A Pentium III 750 MHz CPU combined with a high-powered graphics chip can barely decode a 480P DTV stream. The current crop of high-end processors is necessary to be capable of decoding true HDTV formats such as 720P and 1080I. Software decoding of true HDTV formats demands CPUs with speeds as high as 1 GHz for software-only decoding. Table 9.2 maps the decoding workload.

This solution has the advantage of working with any graphics controller, however the software approach will not work with the huge installed base of lower-speed PCs.

9.3.2 Hardware Assisted Software DTV Decoding

This approach relies on a combination of CPU power and motion-compensation hardware found in graphics chips. There are three components in the hardware assisted software-based solution: a receiver card, a high-powered CPU, and the graphics chip. The receiver card, consisting of a DTV tuner and 8-VSB demodulator, receives the DTV broadcast signal and passes it to the CPU for further processing. The CPU demultiplexes the transport signal and decodes the DTV data using software algorithms. The graphics chip assists the CPU with its on-chip motion compensation and inverse discrete cosine transformation (IDCT) logic.

The approach also strains the CPU and takes up a significant amount of the CPU bandwidth. A Pentium III 550 MHz CPU combined with a high-

Table 9.3 Hardware Assisted DTV Decoding Workload

DTV Hardware	Over-the-air reception VSB demodulation
System CPU	ATSC demultiplex Partial MPEG-2 decoding AC3 decoding Applications overhead
Graphics Hardware	Screen display IDCT Motion compensation

powered graphics chip can decode a 1080I DTV stream. The current crop of high-end processors is necessary to be capable of decoding true HDTV formats such as 720P and 1080I.

Like the software-only approach, this approach will not work with the huge installed base of lower-speed PCs. The installed base of PCs with speeds less than 400 MHz is well over 200 million. These PCs do not have the bandwidth to decode even the 480P transport stream. Dedicated hardware DTV decoding is the only solution for these PCs. (See Table 9.3.)

9.3.3 Hardware DTV Decoding

In this implementation, a dedicated hardware decoder is used for enabling DTV on a PC. A typical system is illustrated in Figure 9.1. The DTV decoder does not rely on the CPU or the graphics chip for decoding the DTV stream. There is very little drain on CPU bandwidth as the add-in card decodes the complete MP@HL MPEG stream. All 18 ATSC DTV formats can be supported at full frame rates. A hardware-based implementation will work with most PCs—even a 166 MHz Pentium PC. This makes hardware DTV decoding an ideal solution for the large installed base of PCs. The workload division for this approach is given in Table 9.4.

A DTVPC solution must meet the following criteria for it to be widely accepted among consumers.

- Single-board solution for true HDTV and legacy analog TV reception

- Simple and easy installation

- Consumer price points for mass-market acceptance

- Support of PC screen formats

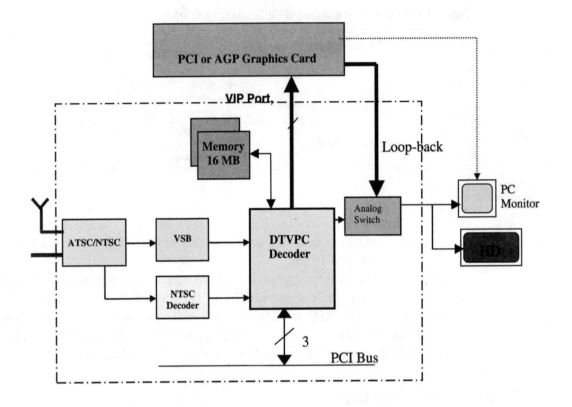

Figure 9.1 Single-board hardware-based DTV decoder.

The DTV card must also support multi-sync monitors and common PC display resolutions such as 1024×768, 800×600, and 640×480. If the board can support DVD playback in addition to ATSC and NTSC, the value proposition for the consumer is significantly enhanced.

The PCI DTVPC tuner card shown in Figure 9.1 is implemented in a single-chip decoder solution. The card has 16 MB of SDRAM memory to support true HDTV display formats (up to 720P and 1080I). An analog loop-back cable connects the DTV card to the graphics subsystem. The analog cable brings the RGB video data and horizontal and vertical sync from the graphics card to the tuner card. The analog multiplexer selects between two different sources: the DTVPC decoder and the graphics card. The DTVPC decoder can be programmed to control the analog switch. Regardless of the incoming video format, this implementation allows full-screen HDTV

Table 9.4 Hardware DTV Decoding Workload

DTV Hardware	Over-the-air reception VSB demodulation ATSC demultiplex Full MPEG-2 decoding Screen display
System CPU	AC3 decoding Applications overhead

broadcast to be displayed on the PC monitor when the decoder video output is selected.

This implementation also allows a video window to be displayed on the graphics desktop. The decoder can simultaneously output CCIR 601 compatible digital video with the high-definition analog video. The CCIR 601 video data may be pumped into the video capture port of the graphics card via the *video interface port* (VIP) of the PCI bus. The graphics card can overlay the incoming video data in a video window on top of the graphics. The viewer can select either full-screen HDTV output or can view the DTV programming in a video window while working on a PC application.

9.4 Intelligent Interactive Devices

As we stated earlier in this chapter, there is a convergence taking place in the consumer electronics (CE) marketplace. This convergence is made possible by the fact that CE devices are becoming more intelligent. The more complex the features that are added to CE appliances, the more processing power is required to allow consumers to easily use these features. This increasing processing within CE appliances has an added benefit in that it becomes easier for manufactures to create appliances that are able to communicate with one another. The following is a list of some of the home appliances that have taken advantage of this device interactivity:

- Television set

- Set top box

- Audio system

- Home control system

- Digital camcorder

- PDA

- CD player

- MP3 player

The PC is also right at home with respect to interactivity. In fact, the PC has lead the way in the development of many of the interconnect standards used by the CE industry today. Serial communications, parallel communications, Internet links, Firewire, USB, have all found their way into the PC. This makes the PC an ideal member of this "connected" family, particularly with its significant processing power, which allows the PC to be a control hub for the entire environment.

9.4.1 The Home as a Digital Hub

We are now seeing the evolution of the digital home. The PC does and will continue to play a central role in this evolution. Increasingly we will bring more and more functionality into the digital home as we improve and add to our entertainment choices, add more Internet service facilities, improve the convenience of our existing services, and unable new education services.

DTVPC solutions will be on the forefront of the digital video evolution just as the PC is at the forefront of the digital home evolution. Exciting new technologies are enhancing our enhancing our enjoyment of television and movies, improving our education, allowing us greater access to more information, and providing more services directly to our digital home.

Interactive Videoconferencing

Jerry C. Whitaker, Editor

10.1 Introduction

With desktop computers nearly as ubiquitous in business these days as telephones, the time has arrived for the next big push in telecommunications—desktop interactive videoconferencing. Interaction via video has been used successfully for many years to permit groups of persons to communicate from widely-distant locations. Such efforts have usually required some degree of advance planning and specialized equipment ranging from custom-built fiber or coax services to satellite links. These types of applications will certainly continue to grow, as the need to communicate on matters of business expands. The real explosion in videoconferencing, however, will come when three criteria are met:

- Little—if any—advance planning is needed

- No special communications links need be installed to participate in a conference

- Participants can do it from their offices

The real promise of interactive videoconferencing is to make it as convenient and accessible as a telephone call. While that day is not here just yet, it will come.

10.2 Infrastructure Issues

Clearly, desktop videoconferencing is the next big thing for business. The challenge of videoconferencing is not the video itself. Required equipment includes one or more cameras, microphones, displays, and the equipment to control them. All of this is important in any type of video conference, and strides are being made in this area all the time. Automatic control of camera movements and intelligent audio switching are just two of the innovative technologies that have been developed to a high level of sophistication. The primary remaining challenge is getting the signal to and from the video equipment on each end. To accomplish this, we need to look toward computer networks, common carriers, and—quite possibly—digital television signals. Although DTV signals are by design one-way, the bandwidth requirements of the *back-channel* are usually modest in comparison to the primary "broadcast" element. As such, the back-channel data can be carried over any common network connection, including a dial-up line.

10.2.1 Full Motion Video

The ideal for any videoconferencing system is the highest possible video quality, in terms of both frame rate and image quality. Some market analysts estimate that as much as 70% of the videoconferencing market falls into this category, partly as a result of the nature of the application, but also because of the fact that most people have been inadvertently trained to expect television quality when they use videoconferencing equipment. This is due—of course—to the prevalence of TV in our society.

Despite the obvious focus on video, audio quality is usually the most important element of a videoconferencing system because it is the audio that conveys most of the information. The whole idea of a videoconference is to facilitate communication among participants. Research has shown that if the audio is clear and audible, the next area on which users focus is the video quality.

A popular myth in the videoconferencing industry is that all systems that conform to established videoconferencing standards (H.320, H.323, and H.324) offer pretty much the same video quality. There may be, however, a significant difference in the video quality of these services as offered by different vendors. Sometimes this difference is a function of the cost of the system, and sometimes it is price-independent—meaning that two vendors could be charging the same price for their product but have very different video quality.

Image quality for a videoconferencing system is largely determined by the quality of the codec being used and the bandwidth allocated to the session. Moreover, much of the codec quality is determined not simply by how much hardware is being used and how fast it is, but by the sophistication of the algorithms that run on the hardware.

10.2.2 Applicable Standards

Standards for hardware and software are the keys to interactive desktop videoconferencing. The whole idea of making videoconferencing convenient depends on systems from different vendors working together. It would—of course—make no sense to buy a telephone on which you could call only certain phones. Although this is certainly an over-simplification of the interface difficulties involved, the point is that the success of videoconferencing will be driven not so much by the attributes of one device over another, but more by which systems will work together.

It is fair to point out, however, that just because there is an industry standard, it does not mean that there will be complete compatibility; problems do arise in the field. Still, standardization must be the driving force behind new product development.

The basic videoconferencing standards developed by the ITU include the following:

- **H.320**: This was the initial videoconferencing standard issued in 1992. It deals only with ISDN-based systems.

- **H.323**: A LAN-based videoconferencing specification. Because of the great progress made recently in moving video over LANs, H.323 is seen as an important component in video-based communications within corporations. Proponents have predicted the replacement of text-based e-mail with video-based e-mail in the not too distant future.

- **H.324**: A standard designed for use with conventional twisted-pair analog lines using 28.8 kbits/s (or faster) modems and sophisticated compression techniques. H.324 was released in 1996 with the hope that it would ensure that all videoconferencing systems, whether stand-alone or computer-based, would be able to talk to each other.

H.324 merits some additional discussion. The standard specifies a common method for video, voice, and data to be shared simultaneously over a single analog line. Audio data is compressed to 6 kbits/s with the remaining 22 kbits/s (for a 28.8 kbits/s modem) allocated to the video signal. In recognition of the importance of the audio link in a videoconference, if bandwidth problems arise during a session, the audio is protected and the video is allowed to degrade. Because most link problems are transient in nature, under this scenario, the picture would begin to break up or otherwise deteriorate, but normal conversations would continue. After the full bandwidth of the channel was restored, the video would return to normal.

Within each of the videoconferencing standards are a number of related ITU standards that specify key operating parameters. For example, under the umbrella of H.324 are the following:

- H.263, which relates to video coding

- H.245, a protocol package

- H.223, which covers the multiplexing video, audio, and user data

- T.120, which addresses user data, whiteboard, file transfer, and other functions

This so-called standards-within-a-standard approach facilitates rapid product development and ease of interconnection among different systems and applications.

10.3 Desktop Systems—Hardware and Software

Implementation of a videoconferencing system on the desktop is a party to which both hardware and software must be invited. There are obvious hardware peripherals that must be present for the system to work, such as a camera, microphone, speaker, and network connection (or modem). The key element of the system is the codec, which can be implemented either in hardware or software. Not surprisingly, the hardware approach is typically faster but more expensive. Software-only codecs, however, have come a long way, and the continuing speed improvements in personal computer CPUs capitalize upon these improvements.

The attractiveness of a software codec is that it can run on any machine, and there is no add-on hardware to purchase. Software codecs also can be updated easily by changing the software driver. Naturally, the performance of the software codec is a function of the type and speed of the processor.

The software aspect of videoconferencing also includes considerations for general office application programs. Interface capabilities with such common desktop apps as Microsoft PowerPoint, Excel, and NetMeeting offer a host of user benefits. For example, a PowerPoint slideshow can be viewed during the videoconference and changes made in real-time. NetMeeting and similar programs that comply with the T.120 standard for data collaboration allow interoperability among distant locations and different vendors.

Control and coordination of a videoconference has been made much easier thanks to new, smart peripheral and supervisory systems. Items such as touch-screens, auto-tracking cameras, and participant responses systems permit users to focus on communicating ideas rather than on producing the videoconference. For example, students may electronically "raise their hand" by pressing a button on a microphone. This notifies the instructor that someone has a question and directs the camera to zoom-in on the student. Through such face-to-face interaction, students and instructors can build relationships that make distance learning more responsive and effective.

10.3.1 Small Group-vs.-Large Group Systems

The nature of a desktop videoconferencing system is that it is usually restricted to one or perhaps two individuals at each terminal. For larger groups, roll-around portable systems and theater-type installations are options. Until recently, the theater-type or dedicated videoconferencing center defined what a video conference was all about: bringing together large groups of people in distant locations. While these applications still exist in large numbers, the primary growth of videoconferencing involves smaller groups.

Portable, roll-around systems offer the benefit of easy setup for small conference rooms and group workspaces. Such packaged systems fill the growing need for group collaboration among distant points.

10.3.2 System Implementations

Although any given videoconferencing session must be designed to meet the particular needs of the subject matter, there are three basic topologies that can be applied to the application:

- *Point-to-point*, the simplest arrangement where two individuals are interconnected in real-time (Figure 10.1*a*).

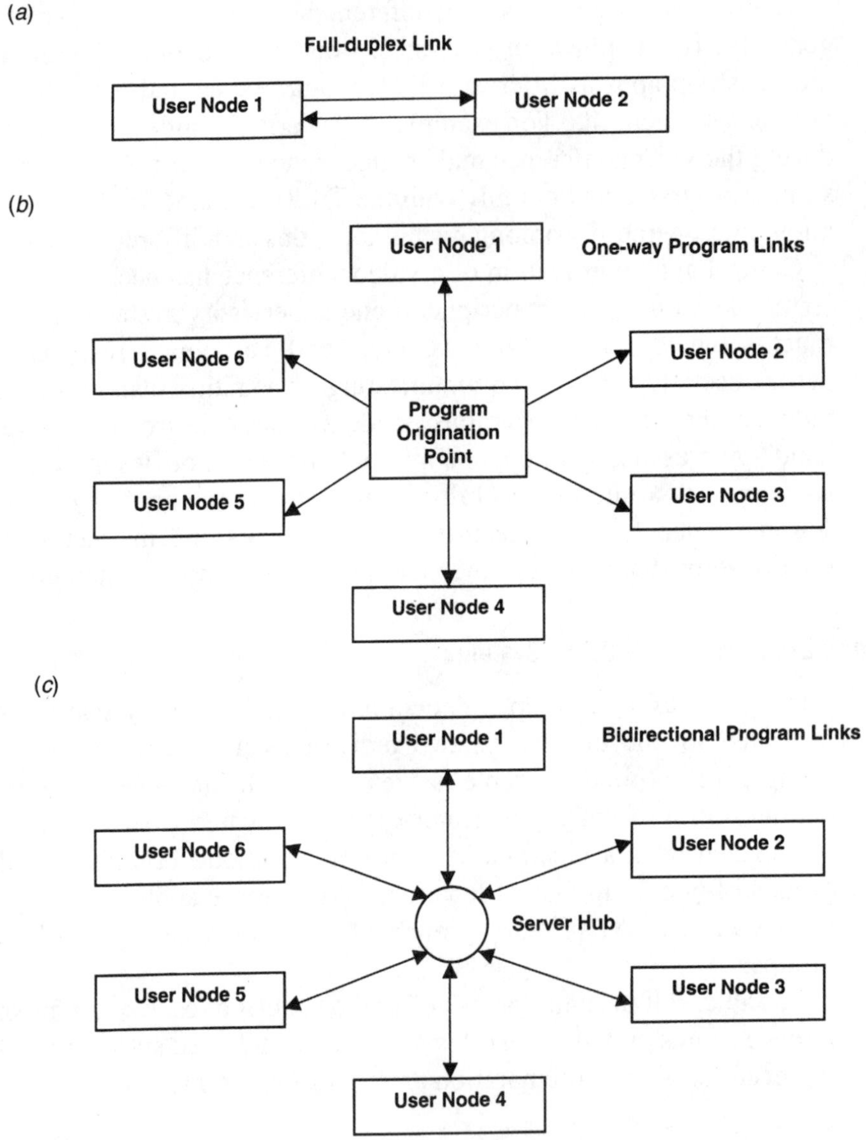

(*a*)

Full-duplex Link

```
┌──────────────┐                    ┌──────────────┐
│  User Node 1 │◄──────────────────►│  User Node 2 │
└──────────────┘                    └──────────────┘
```

(*b*)

```
                    ┌──────────────┐
                    │  User Node 1 │      One-way Program Links
                    └──────────────┘
                           ▲
┌──────────────┐           │              ┌──────────────┐
│  User Node 6 │◄────┐  ┌──────────┐  ┌──►│  User Node 2 │
└──────────────┘     └──│ Program  │──┘   └──────────────┘
                        │Origination│
┌──────────────┐     ┌──│  Point   │──┐   ┌──────────────┐
│  User Node 5 │◄────┘  └──────────┘  └──►│  User Node 3 │
└──────────────┘           │              └──────────────┘
                           ▼
                    ┌──────────────┐
                    │  User Node 4 │
                    └──────────────┘
```

(*c*)

```
                    ┌──────────────┐
                    │  User Node 1 │      Bidirectional Program Links
                    └──────────────┘
                           ▲│
┌──────────────┐           ││             ┌──────────────┐
│  User Node 6 │◄────┐   ╭────╮   ┌──────►│  User Node 2 │
└──────────────┘     └──►│    │◄──┘       └──────────────┘
                         │    │  Server Hub
┌──────────────┐     ┌──►│    │◄──┐       ┌──────────────┐
│  User Node 5 │◄────┘   ╰────╯   └──────►│  User Node 3 │
└──────────────┘           │▲             └──────────────┘
                           ▼│
                    ┌──────────────┐
                    │  User Node 4 │
                    └──────────────┘
```

Figure 10.1 Primary videoconferencing modes: (*a*) point-to-point, (*b*) broadcast, (*c*) multicast.

- *Broadcast*, with a single origination point and multiple receiving points (Figure 10.1*b)*

- *Multicast*, where all (or at least some) of the participating individuals can communicate with each other and/or with the group as a whole (Figure 10.1*c*).

10.3.3 Making Connections

Advances in codec technology have largely eliminated the need for dedicated leased telephone lines. POTS (*plain old telephone service*) twisted-pair lines can deliver video frame rates of 6 to 15 frames per second (f/s), which is sufficient for applications where there is little movement by participants. This situation, of course, essentially describes desktop videoconferencing. With an ISDN line and appropriate hardware, frame rates of 20 to 30 f/s can be realized. *Digital subscriber link* (DSL) service permits full-motion video with little compromises in quality.

The wide variety of interconnection systems used in business today has led videoconference system designers to offer a range of connection options. Some of the more common options include LAN and WAN (wide-area network) based on one or more of the following:

- POTS

- ISDN

- DSL

- Ethernet

- Token Ring

- ATM (asynchronous transfer mode)

- Internet (which, of course, consists of elements of all the above)

Not surprisingly, interconnecting between divergent systems can be problematic. Conversion devices and systems are available, however, to facilitate the transfer of data from one system to another. The widespread use of Internet Protocol has made this technology the prime contender for desktop videoconferencing, and some implementations of conventional, theater-style, conferencing, as well.

Of the modem-function connection options available today, ISDN offers perhaps the best package when measured by the yardsticks of price, performance, availability, and flexibility. ISDN utilizes existing twisted-pair phone lines. ISDN is purely digital; it is not translated into audio-frequency tones as used by analog modems. This makes ISDN more reliable than POTS, allows much higher bandwidth, and greatly reduces the susceptibility to noise. DSL and ISDN share a number of features, although DSL offers a considerable improvement in bandwidth.

End-user equipment includes a *network terminator* (NT) and any number of *terminal adapters*. The NT acts as the bridge between the phone company network and the end user. The terminal adapter is the final piece of equipment used for the application, such as a DSL phone, fax machine, or DSL modem. A terminal adapter refers to any device that is used to generate traffic on a line.

An ISDN connection to the Internet, or any other network for that matter, is a dial-up process. In contrast, DSL systems are usually configured in an "always connected" mode. Optimized for use on the Internet, DSL also provides for dynamic IP address assignment.

11
Glossary

11.1 Terms Employed

For the purposes of the ATSC digital television standard, the following definition of terms apply [1–6]:

16 VSB Vestigial sideband modulation with 16 discrete amplitude levels.

8 VSB Vestigial sideband modulation with 8 discrete amplitude levels.

access unit A coded representation of a presentation unit. In the case of audio, an access unit is the coded representation of an audio frame. In the case of video, an access unit includes all the coded data for a picture, and any *stuffing* that follows it, up to but not including the start of the next access unit.

anchor frame A video frame that is used for prediction. *I*-frames and *P*-frames are generally used as anchor frames, but *B*-frames are never anchor frames.

asynchronous transfer mode (ATM) A digital signal protocol for efficient transport of both constant-rate and variable-rate information in broadband digital networks. The ATM digital stream consists of fixed-length packets called *cells*, each containing 53 8-bit bytes (a 5-byte header and a 48-byte information payload).

bidirectional pictures (B-pictures or **B-frames)** Pictures that use both future and past pictures as a reference. This technique is termed *bidirectional prediction*. *B*-pictures provide the most compression. *B*-pictures do not propagate coding errors as they are never used as a reference.

bit rate The rate at which the compressed bit stream is delivered from the channel to the input of a decoder.

block An 8-by-8 array of DCT coefficients representing luminance or chrominance information.

byte-aligned A bit stream operational condition. A bit in a coded bit stream is byte-aligned if its position is a multiple of 8-bits from the first bit in the stream.

channel A medium that stores or transports a digital television stream.

coded representation A data element as represented in its encoded form.

compression The reduction in the number of bits used to represent an item of data.

constant bit rate The operating condition where the bit rate is constant from start to finish of the compressed bit stream.

conventional definition television (CDTV) This term is used to signify the *analog* NTSC television system as defined in ITU-R Rec. 470. (*See also standard definition television* and ITU-R Rec. 1125.)

CRC Cyclic redundancy check, an algorithm used to verify the correctness of data.

decoded stream The decoded reconstruction of a compressed bit stream.

decoder An embodiment of a decoding process.

decoding (process) The process defined in the ATSC digital television standard that reads an input coded bit stream and outputs decoded pictures or audio samples.

decoding time-stamp (DTS) A field that may be present in a PES packet header which indicates the time that an access unit is decoded in the system target decoder.

D-frame A frame coded according to an MPEG-1 mode that uses dc coefficients only.

DHTML (dynamic HTML) A term used by some vendors to describe the combination of HTML, style sheets, and scripts that enable the animation of web pages.

DOM (document object model) A platform- and language-neutral interface that allows programs and scripts to dynamically access and update the content, structure, and style of documents. The document can be further processed and the results of that processing can be incorporated back into the presented page.

digital storage media (DSM) A digital storage or transmission device or system.

discrete cosine transform (DCT) A mathematical transform that can be perfectly undone and which is useful in image compression.

editing A process by which one or more compressed bit streams are manipulated to produce a new compressed bit stream. Conforming edited bit streams are understood to meet the requirements defined in the ATSC digital television standard.

elementary stream (ES) A generic term for one of the coded video, coded audio, or other coded bit streams. One elementary stream is carried in a sequence of PES packets.

elementary stream clock reference (ESCR) A time stamp in the PES stream from which decoders of PES streams may derive timing.

encoder An embodiment of an encoding process.

encoding (process) A process that reads a stream of input pictures or audio samples and produces a valid coded bit stream as defined in the ATSC digital television standard.

entitlement control message (ECM) Private conditional access information that specifies control words and possibly other stream-specific, scrambling, and/or control parameters.

entitlement management message (EMM) Private conditional access information that specifies the authorization level or the services of specific decoders. They may be addressed to single decoders or groups of decoders.

entropy coding The process of variable-length lossless coding of the digital representation of a signal to reduce redundancy.

entry point A point in a coded bit stream after which a decoder can become properly initialized and commence syntactically correct decoding. The first transmitted picture after an entry point is either an *I*-picture or a *P*-

picture. If the first transmitted picture is not an *I*-picture, the decoder may produce one or more pictures during acquisition.

event A collection of elementary streams with a common time base, an associated start time, and an associated end time.

field For an interlaced video signal, a *field* is the assembly of alternate lines of a frame. Therefore, an interlaced frame is composed of two fields, a top field and a bottom field.

frame Lines of spatial information of a video signal. For progressive video, these lines contain samples starting from one time instant and continuing through successive lines to the bottom of the frame. For interlaced video, a frame consists of two fields, a top field and a bottom field. One of these fields will commence one field later than the other.

group of pictures (GOP) One or more pictures in sequence.

high-definition television (HDTV) An imaging system with a resolution of approximately twice that of conventional television in both the horizontal (H) and vertical (V) dimensions, and a picture aspect ratio (H × V) of 16:9. ITU-R Rec. 1125 further defines "HDTV quality" as the delivery of a television picture that is subjectively identical with the interlaced HDTV studio standard.

high level A range of allowed picture parameters defined by the MPEG-2 video coding specification that corresponds to high-definition television.

HTML (hypertext markup language) A collection of tags typically used in the development of Web pages.

HTTP (hypertext transfer protocol) A set of instructions for communication between a server and a World Wide Web client.

Huffman coding A type of source coding that uses codes of different lengths to represent symbols which have unequal likelihood of occurrence.

intra-coded pictures (*I*-pictures or *I*-frames) Pictures that are coded using information present only in the picture itself and not depending on information from other pictures. *I*-pictures provide a mechanism for random access into the compressed video data. *I*-pictures employ transform coding of the pixel blocks and provide only moderate compression.

layer One of the levels in the data hierarchy of the DTV video and system specifications.

level A range of allowed picture parameters and combinations of picture parameters.

macroblock In the advanced television system, a macroblock consists of four blocks of luminance and one each C_r and C_b block.

main level A range of allowed picture parameters defined by the MPEG-2 video coding specification, with maximum resolution equivalent to ITU-R Rec. 601.

main profile A subset of the syntax of the MPEG-2 video coding specification that is supported over a large range of applications.

MIME (multipart/signed, multipart/encrypted content-types) A protocol for allowing e-mail messages to contain various types of media (text, audio, video, images, etc.).

motion vector A pair of numbers that represent the vertical and horizontal displacement of a region of a reference picture for prediction purposes.

MPEG Standards developed by the ISO/IEC JTC1/SC29 WG11, *Moving Picture Experts Group*. MPEG may also refer to the Group itself.

MPEG-1 ISO/IEC standards 11172-1 (Systems), 11172-2 (Video), 11172-3 (Audio), 11172-4 (Compliance Testing), and 11172-5 (Technical Report).

MPEG-2 ISO/IEC standards 13818-1 (Systems), 13818-2 (Video), 13818-3 (Audio), and 13818-4 (Compliance).

pack A header followed by zero or more packets; a layer in the ATSC DTV system coding syntax.

packet A header followed by a number of contiguous bytes from an elementary data stream; a layer in the ATSC DTV system coding syntax.

packet data Contiguous bytes of data from an elementary data stream present in the packet.

packet identifier (PID) A unique integer value used to associate elementary streams of a program in a single or multi-program transport stream.

padding A method to adjust the average length of an audio frame in time to the duration of the corresponding PCM samples by continuously adding a slot to the audio frame.

payload The bytes that follow the header byte in a packet. The transport stream packet header and adaptation fields are not payload.

PES packet The data structure used to carry elementary stream data. It consists of a packet header followed by PES packet payload.

PES stream A stream of PES packets, all of whose payloads consist of data from a single elementary stream, and all of which have the same stream identification.

picture Source, coded, or reconstructed image data. A source or reconstructed picture consists of three rectangular matrices representing the luminance and two chrominance signals.

pixel "Picture element" or "pel." A pixel is a digital sample of the color intensity values of a picture at a single point.

predicted pictures (*P*-pictures or *P*-frames) Pictures that are coded with respect to the nearest *previous I* or *P*-picture. This technique is termed *forward prediction*. *P*-pictures provide more compression than *I*-pictures and serve as a reference for future *P*-pictures or *B*-pictures. *P*-pictures can propagate coding errors when *P*-pictures (or *B*-pictures) are predicted from prior *P*-pictures where the prediction is flawed.

presentation time-stamp (PTS) A field that may be present in a PES packet header that indicates the time that a presentation unit is presented in the system target decoder.

presentation unit (PU) A decoded audio access unit or a decoded picture.

profile A defined subset of the syntax specified in the MPEG-2 video coding specification.

program A collection of program elements. Program elements may be elementary streams. Program elements need not have any defined time base; those that do have a common time base and are intended for synchronized presentation.

program clock reference (PCR) A time stamp in the transport stream from which decoder timing is derived.

program element A generic term for one of the elementary streams or other data streams that may be included in the program.

program specific information (PSI) Normative data that is necessary for the demultiplexing of transport streams and the successful regeneration of programs.

quantizer A processing step that intentionally reduces the precision of DCT coefficients

random access The process of beginning to read and decode the coded bit stream at an arbitrary point.

scrambling The alteration of the characteristics of a video, audio, or coded data stream in order to prevent unauthorized reception of the information in a clear form.

slice A series of consecutive macroblocks.

source stream A single, non-multiplexed stream of samples before compression coding.

splicing The concatenation performed on the system level or two different elementary streams. It is understood that the resulting stream must conform totally to the ATSC digital television standard.

standard definition television (SDTV) This term is used to signify a *digital* television system in which the quality is approximately equivalent to that of NTSC. This equivalent quality may be achieved from pictures sourced at the 4:2:2 level of ITU-R Rec. 601 and subjected to processing as part of bit rate compression. The results should be such that when judged across a representative sample of program material, subjective equivalence with NTSC is achieved. Also called standard digital television.

start codes 32-bit codes embedded in the coded bit stream that are unique. They are used for several purposes, including identifying some of the layers in the coding syntax.

STD input buffer A first-in, first-out buffer at the input of a system target decoder (STD) for storage of compressed data from elementary streams before decoding.

still picture A video sequence containing exactly one coded picture that is intra-coded. This picture has an associated PTS and the presentation time of succeeding pictures, if any, is later than that of the still picture by at least two picture periods.

system clock reference (SCR) A time stamp in the program stream from which decoder timing is derived.

system header A data structure that carries information summarizing the system characteristics of the ATSC digital television standard multiplexed bit stream.

system target decoder (STD) A hypothetical reference model of a decoding process used to describe the semantics of the ATSC digital television standard multiplexed bit stream.

time-stamp A term that indicates the time of a specific action such as the arrival of a byte or the presentation of a presentation unit.

transport stream packet header The leading fields in a transport stream packet.

UHTTP (unidirectional hypertext transfer protocol) A is a simple, robust, one-way resource transfer protocol that is designed to efficiently deliver resource data in a one-way broadcast-only environment. This resource transfer protocol is appropriate for IP multicast over the television vertical blanking interval (IP-VBI), in an IP multicast carried in MPEG-2, or in other unidirectional transport systems.

UUID (universally unique identifier) An identifier that is unique across both space and time, with respect to the space of all UUIDs. Also known as GUID (globally unique identifier).

variable bit rate An operating mode where the bit rate varies with time during the decoding of a compressed bit stream.

video buffering verifier (VBV) A hypothetical decoder that is conceptually connected to the output of an encoder. Its purpose is to provide a constraint on the variability of the data rate that an encoder can produce.

video sequence An element represented by a sequence header, one or more groups of pictures, and an end of sequence code in the data stream.

11.2 Acronyms and Abbreviations

A/D analog to digital converter

ACATS Advisory Committee on Advanced Television Service

AES Audio Engineering Society

ANSI American National Standards Institute

ATEL Advanced Television Evaluation Laboratory

ATM asynchronous transfer mode

ATSC Advanced Television Systems Committee

ATTC Advanced Television Test Center

ATV advanced television

bps bits per second

bslbf bit serial, leftmost bit first

CAT conditional access table

CDT carrier definition table

CDTV conventional definition television

CRC cyclic redundancy check

DCT discrete cosine transform

DSM digital storage media

DSM-CC digital storage media command and control

DTS decoding time-stamp

DVCR digital video cassette recorder

ECM entitlement control message

EMM entitlement management message

ES elementary stream

ESCR elementary stream clock reference

FPLL frequency- and phase-locked loop

GA Grand Alliance

GMT Greenwich mean time

GOP group of pictures

GPS global positioning system

HDTV high-definition television

IEC International Electrotechnical Commission

IRD integrated receiver-decoder

ISO International Organization for Standardization

ITU International Telecommunication Union

JEC Joint Engineering Committee of EIA and NCTA

MCPT multiple carriers per transponder

MMT modulation mode table

MP@HL main profile at high level

MP@ML main profile at main level

MPEG Moving Picture Experts Group

NAB National Association of Broadcasters

NTSC National Television System Committee

NVOD near video on demand

PAL phase alternation each line

PAT program association table

PCR program clock reference

pel pixel

PES packetized elementary stream

PID packet identifier

PMT program map table

PSI program specific information

PTS presentation time stamp

PU presentation unit

SCR system clock reference

SDTV standard definition television

SECAM sequential couleur avec mémoire (sequential color with memory)

SIT satellite information table

SMPTE Society of Motion Picture and Television Engineers

STD system target decoder

TAI international atomic time

TDT transponder data table

TNT transponder name table

TOV threshold of visibility

TS transport stream

UTC universal coordinated time

VBV video buffering verifier

VCN virtual channel number

VCT virtual channel table

11.3 References

1. *ATSC Digital Television Standard,* Doc. A/53, Advanced Television Systems Committee, Washington, D.C., 1996.

2. *Digital Audio Compression (AC-3) Standard,* Doc. A/52, Advanced Television Systems Committee, Washington, D.C., 1996.

3. *Guide to the Use of the ATSC Digital Television Standard,* Doc. A/54, Advanced Television Systems Committee, Washington, D.C., 1996.

4. *Program Guide for Digital Television,* Doc. A/55, Advanced Television Systems Committee, Washington, D.C., 1996.

5. *System Information for Digital Television,* Doc. A/56, Advanced Television Systems Committee, Washington, D.C., 1996.

6. "Advanced Television Enhancement Forum Specification," Draft, Version 1.1r26 updated 2/2/99, ATVEF, Portland, Ore., 1999.

Subject Index

Numerics

2D transform 41
8-VSB 249
16 × 9 format 26
16-VSB 249

A

ac coefficients 41
access control 194
access unit 249
active picture area 60
adaptive differential PCM 89
adaptive PCM 89
addressable section encapsulation 131
advanced audio coding (AAC) 102
Advanced Authoring Format 201
Advanced Television Enhancement Forum 14, 210
Advanced Television Systems Committee 169
Amendment No.1 to ATSC Standard A/65 116
analog-to-digital conversion 10
anchor frame 249
announcement data 214
application 140

application programming interface 17, 162
apt-X100 95
artifacts 82
ASPEC 92
aspect ratio 186
asynchronous data 140
asynchronous datagram 163
asynchronous transfer mode 249
ATSC 169
ATSC T3/S17 specialist group 15
attributes 197
ATV Forum 211
audio bit rates 93
audio compression 86
audio-visual event 140
aural monitoring 35

B

back-channel 213
backward prediction 50
best-efforts network 222
B-frame 61
bidirectional pictures 249
binding 133
bit rate 140, 250
bit-rate-reduction 40
bit-rate-reduction systems 86

Jerry C. Whitaker is President of Technical Press, a consulting company based in the San Jose (CA) area. Mr. Whitaker has been involved in various aspects of the electronics industry for over 25 years, with specialization in communications. Current book titles include the following:

- Editor-in-chief: *Standard Handbook of Video and Television Engineering*, 3rd ed., McGraw-Hill, 2000

- Editor: *Video and Television Engineers' Field Manual*, McGraw-Hill, 2000

- *Video Display Engineering*, McGraw-Hill, 2000

- *Radio Frequency Transmission Systems: Design and Operation*, McGraw-Hill, 1990

- Editor-in-chief, *The Electronics Handbook*, CRC Press, 1996

- *AC Power Systems Handbook*, 2nd ed., CRC Press, 1999

- *Power Vacuum Tubes Handbook*, 2nd ed., CRC Press, 1999

- *The Communications Facility Design Handbook*, CRC Press, 2000

- *The Resource Handbook of Electronics*, CRC Press, 2000

- Co-author, *Television and Audio Handbook for Engineers and Technicians*, McGraw-Hill, 1989

- Co-author, *Communications Receivers*, 3rd ed., McGraw-Hill, 2000

- Co-editor, *Information Age Dictionary*, Intertec Publishing/Bellcore, 1992

Mr. Whitaker has lectured extensively on the topic of electronic systems design, installation, and maintenance. He is the former editorial director and associate publisher of *Broadcast Engineering* and *Video Systems* magazines, and a former radio station chief engineer and television news producer.

Mr. Whitaker is a Fellow of the Society of Broadcast Engineers and an SBE-certified professional broadcast engineer. He is also a fellow of the

Society of Motion Picture and Television Engineers, and a member of the Institute of Electrical and Electronics Engineers.

Mr. Whitaker has twice received a Jesse H. Neal Award *Certificate of Merit* from the Association of Business Publishers for editorial excellence. He has also been recognized as *Educator of the Year* by the Society of Broadcast Engineers.

Mr. Whitaker resides in Morgan Hill, California.

On-Line Updates

Additional updates relating to DTV in general, and this book in particular, can be found at the *Standard Handbook of Video and Television Engineering* web site:

www.tvhandbook.com

The tvhandbook.com web site supports the professional video community with news, updates, and product information relating to the broadcast, post production, and business/industrial applications of digital video.

Check the site regularly for news, updated chapters, and special events related to video engineering. The technologies encompassed by *Interactive TV Demystified* are changing rapidly, with new standards proposed and adopted each month. Changing market conditions and regulatory issues are adding to the rapid flow of news and information in this area.

Specific services found at **www.tvhandbook.com** include:

- **Video Technology News**. News reports and technical articles on the latest developments in digital television, both in the U.S. and around the world. Check in at least once a month to see what's happening in the fast-moving area of digital television.

- **Television Handbook Resource Center**. Check for the latest information on professional and broadcast video systems. The Resource Center provides updates on implementation and standardization efforts, plus links to related web sites.

- **tvhandbook.com Update Port**. Updated material for *DTV: The Revolution in Digital Video* is posted on the site each month. Material available includes updated sections and chapters in areas of rapidly advancing technologies.

- **tvhandbook.com Book Store**. Check to find the latest books on digital video and audio technologies. Direct links to authors and publishers are provided. You can also place secure orders from our on-line bookstore.

In addition to the resources outlined above, detailed information is available on other books in the McGraw-Hill Video/Audio Series.